Memoirs of the American Mathematical Society

Number 171

John Ernest

Charting the operator terrain

Published by the

AMERICAN MATHEMATICAL SOCIETY

Providence, Rhode Island

VOLUME 6 · NUMBER 171 (first of 3 numbers) · JULY 1976

ABSTRACT

The purpose of this memoir is to offer a cartographic procedure for bringing some organizational sense to the prodigious task of exploring and describing the vast and varied terrain of bounded operators on a separable Hilbert space. While we prove (in a certain sense) that the classification of all operators up to unitary equivalence is an essentially unattainable objective, we hope this theory will prove serviceable in colonizing some additional enclaves as well as suggesting other more rugged areas which one might probe in search of fascinating and unusual phenomena. Our cartographic system is essentially an adaptation to single operators of John von Neumann's classification and reduction theory (where the study of rings of operators is reduced to the study of factors.) This single operator version is cast into a mold which exhibits it as a natural generalization of the spectral theorem and the spectral multiplicity theory for normal operators. To each bounded operator we associate a certain set of (equivalence classes of) "factor operators," which is called the quasi-spectrum of the operator. An operator is then determined, up to unitary equivalence, by a triplet consisting of its quasi-spectrum, a finite measure class on that quasi-spectrum, and a "spectral multiplicity function" defined on the measure classes absolutely continuous with respect to the first mentioned measure class. In a certain imprecise and audacious sense, this reduces the unitary equivalence problem for general operators to a (weaker) equivalence problem for "factor operators." The spectral theorem, and the spectral multiplicity theory for normal operators are (very) special cases of this general theory.

AMS(MOS) subject classifications (1970). Primary 47A02, 47A10, 47A60, 47A65, 47B15, 47C10, 47C15; Secondary 22D10, 22D25, 46L05, 46L10.

Library of Congress Cataloging in Publication Data **CIP**

Ernest, John, 1935-
Charting the operator terrain.

(Memoirs of the American Mathematical Society ; no. 171)
"Volume 6."
Bibliography: p.
Includes indexes.
1. Linear operators. 2. Spectral theory (Mathematics) I. Title. II. Series: American Mathematical Society. Memoirs ; 171.
QA3.A57 no.171 [QA329.2] 510'.8s [515'.72]
ISBN 0-8218-1871-6 76-3583

TABLE OF CONTENTS

Dedicated to David Topping

CHARTING THE OPERATOR TERRAIN

INTRODUCTION

The territory of bounded linear operators on a Hilbert space is vast. There are basically two regions of civilization (the normal operators and the compact operators) which are surrounded by various more recent settlements. Beyond this frontier there remains an enormous, almost completely virgin, wilderness. The purpose of this memoir is to offer a cartographic procedure for bringing some organizational sense to the prodigious task of exploring this varied operator terrain. While we will prove (in a certain sense) that the classification of all operators up to unitary equivalence is an essentially unattainable objective, we hope this theory will be serviceable in colonizing some additional enclaves, as well as suggesting other more rugged areas which one might probe in search of some fascinating and unusual phenomena.

Our cartographic system is essentially an adaptation to single operators of John von Neumann's classification and reduction theory, in which the study of general rings of operators is reduced to the study of certain minimal building blocks called factors. Our single operator version is cast into a mold which exhibits it as a natural generalization of the spectral theorem and the spectral multiplicity theory for normal operators. To each bounded operator we associate a certain set of (equivalence classes of) "factor operators," which we call the quasi-spectrum of the operator. This object may be identified with the ordinary spectrum of the operator precisely when the operator is normal. An arbitrary bounded operator is determined, up to unitary equivalence, by a triplet consisting of its quasi-spectrum, a finite measure class on that quasi-spectrum, and a "spectral multiplicity function" defined on the measure classes absolutely continuous with respect to the first mentioned measure class. In a certain imprecise and audacious sense, this reduces the unitary equivalence problem for general operators to a (weaker) equivalence problem for "factor operators." The spectral theorem and the spectral multiplicity theory for normal operators are (very) special cases of this general theory.

About thirty years ago Gelfand and Naimark generalized the spectral theorem for normal operators to a theorem about commutative C^*-algebras. Since that time a beautiful, deep and extensive theory for noncommutative C^*-algebras has developed, as an examination of Jacque Dixmier's basic treatise

Received by the Editors December 6, 1973, and in revised form December 16, 1975.

[36] will verify. A reading of this same treatise will also indicate that the noncommutative theory has become largely divorced from its roots, operator theory proper, perhaps to the detriment of both subjects. In the last few years, however, a number of mathematicians have been using a mixture of operator methods and C^*-algebra techniques and have thus begun to reestablish the deep connections between the two theories: Arveson, Behncke, Bunce, Deddens, Coburn, Douglas, Gonshor, O'Donovan, Okayasu, Pearcy, Topping and Wogen, to mention just a few. The purpose of this memoir is to make as explicit as possible the fact that the general theory of noncommutative C^*-algebras is, in particular, a very extensive theory of non-normal operators. That is to say, we have tried to transplant certain portions of the general C^*-algebra theory back into the rich soil of ordinary operator theory.

This process begins by adopting and adapting a whole glossary from von Neumann algebra theory and formulating the definitions in terms which seem more natural in the operator theoretic context. Here we follow the ideas of George Mackey [94][95], who has adapted the von Neumann algebra classification system to the context of representation theory. While few of the basic notions are new, this still introduces an enormous number of new terms into operator theory and we have tried to minimize the pain by including an index at the end of the memoir.

We consider three basic equivalence relations among operators, which we mention in order of increasing strength: weak equivalence, quasi-equivalence and unitary equivalence. We also introduce the notion of one operator being "weakly contained" in another. A _factor operator_ is defined to be a kind of minimal building block: specifically an operator is a factor operator if it cannot be expressed as a direct sum of two operators such that, no suboperator of one is unitarily equivalent to any suboperator of the other.

The quasi-spectrum $\tilde{T}$ of an operator T is the space of all quasi-equivalence classes of factor operators weakly contained in T. A σ-ring of sets may be specified in $\tilde{T}$ which is called a Borel structure for $\tilde{T}$. The _spectrum_ $\hat{T}$ of an operator T is the space of unitary equivalence classes of irreducible operators weakly contained in T. By a natural identification, $\hat{T}$ may be considered a Borel subset of $\tilde{T}$ and we call an operator _smooth_ if $\hat{T} = \tilde{T}$. In addition to its Borel structure, the spectrum $\hat{T}$ has an interesting (in general non-Hausdorff) topology. The one-dimensional part of $\hat{T}$ has been studied extensively by operator theoreticians under the name "normal approximate spectrum."

Every bounded operator T on a separable Hilbert space has a canonical direct integral decomposition with respect to a finite measure μ defined on its quasi-spectrum $\tilde{T}$. This decomposition, the subject of chapter 3, is our candidate for the spectral theorem for arbitrary operators.

The equivalence problem then goes as follows. Two operators are weakly equivalent if and only if they have the same quasi-spectrum. Two operators are quasi-equivalent if and only if they have the same quasi-spectrum and their "spectral decompositions" determine the same measure class on the quasi-spectrum. Finally we define a generalization of the notion of spectral multiplicity function which is positive real valued (and $+\infty$) rather than cardinal valued. To each operator we associate such a multiplicity function defined on the measures absolutely continuous with respect to the measure on the quasi-dual given by the spectral decomposition of the operator. Finally two operators are unitarily equivalent if and only if they have the same quasi-spectrum, the same measure class and the same multiplicity function. Further to each such multiplicity function there exists a corresponding unitary equivalence class of operators. If T is normal the quasi-spectrum is just the ordinary spectrum, the direct integral decomposition reduces to the ordinary spectral theorem, and the spectral multiplicity theory reduces to the classical theory [77]. (For another set of unitary invariants, see §5.3 of [85].)

If T is a smooth operator the multiplicity function is positive integral (or $+\infty$) valued, i.e., the spectral multiplicity theory takes exactly the same form as in the normal case. The biggest technical snag of the theory (which does not occur in the smooth operator case) is that not every measure class on the quasi-spectrum of an operator, corresponds to an operator by means of the canonical direct integral decomposition. The identification of which measures do arise (in the non-smooth case) is described in the second section of chapter 3.

Admittedly the theory is extremely general and abstract. In chapter 5 we work out a few concrete examples which we hope will suffice to show that this general chart can be used effectively for current explorations in operator theory. In the final section of chapter 5 we conclude by giving our list of eight prime research areas.

The term "operator," with no additional modification, will always mean a bounded linear operator on a separable Hilbert space, finite or infinite dimensional. Actually we will, on a few occasions, consider operators on a non-separable space, but in every such case we shall explicitly indicate that fact.

The theory of this paper is the formulation and adaptation, in operator theoretic terms, of research results obtained, largely in the last twenty years, by an (international) community of mathematicians. As a way of indicating our indebtedness to these individuals I have broken with tradition: the bibliography lists people with names and not just references with initials.

This memoir, both in its point of view as well as its specifics, is largely an operator theoretic version of the approach of George Mackey to representation theory. A certain incurable punster in our ranks would list the subject as a branch of "Mackeymatics".

We are also deeply indebted to Jacques Dixmier, not only for many specific results, but also for his two major treatises [36] [37] without which this paper could not have been written. The reader is hereby warned that reading this paper is almost impossible unless both of these books are within hands reach.

The basic idea for this work has been anticipated by William Arveson, with whom we have had many useful conversations. His paper [3] already indicates many of the fundamental connections between C^*-algebra theory and operators. In the concluding paragraph of that work Arveson clearly suggests a theory similar to the one presented here, at least in the smooth case. Further he in fact outlines the decomposition theory in section 2.3 of [6], again in the smooth case. We have also benefited from conversations with Paul Halmos and the other operator theoreticians at the University of Indiana, during a sabbatical visit there in the spring of 1971.

The complete typing and lay-out of this memoir has been expertly done by Delores Brannon, and I am deeply grateful for the time and care she has taken for this task. It is her original typed manuscript, reduced somewhat in size, that is reproduced here. Louise Kraus has made the bibliography considerably more useful by adding the Mathematical Review references given there. Wai Fong Kwok, who has helped in the editing and proof reading of this final version, has also made some useful mathematical contributions. Finally we wish to acknowledge and thank the National Science Foundation for their support during the research and preparation of this memoir.

The author would very much appreciate receiving comments on the subject matter of this treatise, as well as reprints and preprints of the reader's work in operator theory, C^*-algebras or von Neumann algebras.

Chapter 1

EQUIVALENCE RELATIONS AND CLASSIFICATION THEORY

1. Quasi-equivalence

DEFINITION 1.1. Two operators S and T, acting on Hilbert spaces $\mathcal{H}(S)$ and $\mathcal{H}(T)$ respectively, are unitarily equivalent (denoted $S \simeq T$) if there exists an isometrical isomorphism U of $\mathcal{H}(S)$ onto $\mathcal{H}(T)$ such that $S = U^{-1}TU$.

DEFINITION 1.2. If T is an operator acting on a Hilbert space $\mathcal{H}(T)$, a reducing subspace for T is a closed subspace $\mathcal{H}_o$ of $\mathcal{H}(T)$ ($\dim \mathcal{H}_o \geq 1$) which is invariant under both T and T^*.

If S is an operator which is equal to the restriction of T to a reducing subspace $\mathcal{H}(S)$ for T, we say S is a suboperator of T, denoted $S \leq T$. We say T is irreducible if it has no proper suboperators.

We shall use the notation $S \lesssim T$ to denote the fact that S is unitarily equivalent to a suboperator of T.

The notion of suboperator will be basic throughout this treatise in that the classification of an operator will be given primarily in terms of the structure of its collection of suboperators. We begin with an operator theoretic analogue of the Schröder-Bernstein theorem of set theory (cf. theorem 1.1 of [95] and test problem 3 of [108]).

THEOREM 1.3. Let S and T denote two operators such that $S \lesssim T$ and $T \lesssim S$. Then $S \simeq T$.

PROOF. Let U denote the isometric linear transformation mapping the Hilbert space $\mathcal{H}(S)$ onto $U\mathcal{H}(S)$, a closed subspace of $\mathcal{H}(T)$, such that the restriction of T to $U\mathcal{H}(S)$ is equal to USU^{-1}. Symmetrically let V denote the isometric linear transformation mapping $\mathcal{H}(T)$ onto $V\mathcal{H}(T)$, a closed subspace of $\mathcal{H}(S)$ such that the restriction of S to $V\mathcal{H}(T)$ is equal to VTV^{-1}.

We define a sequence of mutually orthogonal reducing subspaces for S, as follows: Let

$$\begin{aligned} \mathcal{H}_{2k-1} &= (VU)^{k-1}\mathcal{H}(S) - (VU)^{k-1}V\mathcal{H}(T) \\ \mathcal{H}_{2k} &= (VU)^{k-1}V\mathcal{H}(T) - (VU)^{k}\mathcal{H}(S) \end{aligned}$$

for $k = 1,2,\ldots$. (Here $\mathcal{M} - \mathcal{N}$ denotes the orthogonal complement of $\mathcal{N}$ in $\mathcal{M}$, where $\mathcal{N}$ is a closed subspace of $\mathcal{M}$.) Let

$$\mathcal{H}_0 = \mathcal{H}(S) - \sum_{k=1}^{\infty} \oplus \mathcal{H}_k = \bigcap_{k=1}^{\infty} (VU)^k V\mathcal{H}(T).$$

Symmetrically we define a sequence of mutually orthogonal reducing subspaces for T, as follows: Let

$$\begin{aligned} \mathcal{K}_{2k-1} &= (UV)^{k-1}\mathcal{H}(T) - (UV)^{k-1}U\mathcal{H}(S) \\ \mathcal{K}_{2k} &= (UV)^{k-1}U\mathcal{H}(S) - (UV)^{k}\mathcal{H}(T) \end{aligned}$$

for $k = 1,2,\ldots$. Let

$$\mathcal{K}_0 = \mathcal{H}(T) - \sum_{k=1}^{\infty} \oplus \mathcal{K}_k = \bigcap_{k=1}^{\infty} (UV)^k \mathcal{H}(T).$$

Note that U maps $\mathcal{H}_{2k-1}$ onto $\mathcal{K}_{2k}$ for $k = 1,2,\ldots,$ as well as mapping $\mathcal{H}_0$ onto $\mathcal{K}_0$. Further V^{-1} maps $\mathcal{H}_{2k}$ onto $\mathcal{K}_{2k-1}$ for $k = 1,2,\ldots$. We define a linear isometry W of $\mathcal{H}(S) = \sum_{k=0}^{\infty} \oplus \mathcal{H}_k$ onto $\mathcal{H}(T) = \sum_{k=0}^{\infty} \oplus \mathcal{K}_k$ by letting $W\psi = U\psi$ when $\psi \in \mathcal{H}_k$ for k odd or $k = 0$, and letting $W\psi = V^{-1}\psi$ when $\psi \in \mathcal{H}_{2k}$, $k = 1,2,\ldots$. Finally it is easy to verify that $T = WSW^{-1}$.

DEFINITION 1.4. A linear transformation A will be called an intertwining transformation for two operators S and T providing A maps $\mathcal{H}(S)$ into $\mathcal{H}(T)$ and $AS = TA$ and $AS^* = T^*A$.

THEOREM 1.5. (Schur's lemma, cf. Theorem 1.2 of [95].) Let A be an intertwining transformation for the operators S and T. Then $\ker A$ and $\overline{\operatorname{ran} A}$ are reducing subspaces of S and T respectively and the restriction S_o of S to $(\ker A)^\perp$ is unitarily equivalent to the restriction T_o of T to $\overline{\operatorname{ran} A}$.

PROOF. It is easy to verify that $\ker A$ is a reducing space for S and that $\overline{\operatorname{ran} A}$ is a reducing space for T.

We next consider the polar decomposition $A = UH$ of A where $H = (A^*A)^{\frac{1}{2}}$ is a positive operator on $\mathcal{H}(S)$ whose range is dense in $(\ker A)^\perp$ and U is a linear isometry mapping $(\ker A)^\perp$ onto $\overline{\operatorname{ran} A}$ (cf. problem 105 of [78]). Since A intertwines S and T it is easy to verify that $H^2 = A^*A$ commutes with S_o, as does every polynomial in H^2. Thus every operator in the abelian C^*-algebra generated by H^2 commutes with S_o. Hence by proposition 1.6.1 of [36] H commutes with S_o. Hence

$$AS_o = UHS_o = T_oUH = US_oH$$

or

$$(US_o - T_oU)H = 0.$$

But the range of H is dense in $(\ker A)^\perp$, the domain of both U and S_o. Thus $US_o = T_oU$.

COROLLARY 1.6. Suppose S and T are nonzero operators such that $S \lesssim T$ and further that $T = \sum_i \oplus T_i$ is a direct sum of other operators T_i. Then for some i, S has a suboperator unitarily equivalent to a suboperator of T_i.

PROOF. Let T_o denote a suboperator of T which is unitarily equivalent to S and let F denote the projection of $\mathcal{H}(T)$ onto the domain of T_o. For each i, let E_i denote the projection of $\mathcal{H}(T)$ onto the domain of T_i. Since F is nonzero and the E_i are mutually orthogonal and $\sum_i E_i = I$, we have that, for at least one i, the restriction A of E_i to the range of F is nonzero. It is easy to verify that A is an intertwining transformation for T_o and T_i.

DEFINITION 1.7. If S and T are two operators we say they are disjoint (denoted $S \perp T$) if no suboperator of S is unitarily equivalent to any suboperator of T.

COROLLARY 1.8. Two operators are disjoint if and only if they do not admit a nonzero intertwining transformation.

COROLLARY 1.9. If T is an operator which is a direct sum, $T = \sum_i \oplus T_i$, of other operators T_i, then an operator S is disjoint from T if and only if S is disjoint from T_i for every i.

DEFINITION 1.10. We say an operator S covers an operator T (denoted $S \succ T$) if no suboperator of T is disjoint from S. We say S is quasi-equivalent to T (denoted $S \approx T$) if S covers T and T covers S.

The terminology "quasi-equivalence is slightly misleading as this is a bonafide equivalence relation. The notation $\approx$ is consistent with that used in representation theory by Dixmier [36] but not with that used by Mackey [94] [95]. The Mackey notation, $\sim$, will be used in this monograph for another still weaker equivalence relation among operators (definition 1.46).

The notation $\textcircled{n} S$ denotes the direct sum of n copies of the operator S, in some specific concrete manner, where n is a cardinal number. To be canonical we can let $\textcircled{n} S = S \otimes I$, where I is the identity operator on an n-dimensional space. Our next result indicates, intuitively, that quasi-equivalent operators are the same except for "differences in multiplicity."

PROPOSITION 1.11. If S and T are operators then $S \approx T$ if and only if $\textcircled{\infty} S \simeq \textcircled{\infty} T$.

PROOF. First note that for any operator S, $S \approx \textcircled{\infty} S$. Clearly $\textcircled{\infty} S \} S$ and corollary 1.9 implies $S \} \textcircled{\infty} S$. Thus if $\textcircled{\infty} S \simeq \textcircled{\infty} T$ we have $S \approx \textcircled{\infty} S \approx \textcircled{\infty} T \approx T$.

Next suppose $S \approx T$. Since $T \} S$, S is not disjoint from T and hence admits a suboperator which is unitarily equivalent to a suboperator of T. Using Zorn's lemma, there exist a maximal family $\{S_i\}$ of suboperators S_i of S for which each S_i is unitarily equivalent to a suboperator of T and such that spaces $\mathcal{H}(S_i)$ on which the operators S_i act are mutually orthogonal. Furthermore $\mathcal{H}(S) = \sum_i \oplus \mathcal{H}(S_i)$, for otherwise S would admit a suboperator disjoint from T, contradicting the fact that $T \} S$. Thus $S = \sum_{i=1}^{n} \oplus S_i$ where n is some cardinal, $1 \leq n \leq \aleph_0$. Further $S_i \lesssim T$

implies

$$S = \sum_{i=1}^{n} \oplus S_i \lesssim \textcircled{n}T \lesssim \textcircled{\infty}T.$$

Similarly $T \lesssim \textcircled{\infty}S$. Thus $\textcircled{\infty}S \lesssim \textcircled{\infty}(\textcircled{\infty}T) \simeq \textcircled{\infty}T$. Similarly $\textcircled{\infty}T \lesssim \textcircled{\infty}S$. Hence by theorem 1.3 we have $\textcircled{\infty}S \simeq \textcircled{\infty}T$.

We mention that the definitions and results of this section, with the exception of proposition 1.11 above, apply equally well to operators acting on nonseparable Hilbert spaces.

2. Von Neumann classification of operators, in operator theoretic terms

In this section we give a classification system for operators, corresponding to von Neumann's terminology for weakly closed $*$ - algebras of operators. However we shall not refer to von Neumann algebras at all in this section, preferring to develop the complete terminology in purely operator theoretic terms, and compiling all the definitions in this one section for ready reference. Where statements are given without proof, their proof is deferred to the next section where we will make the connection with von Neumann algebra theory explicit. This is not to say these proofs could not be rendered in an operator theoretic form, but only that any attempt to develop the theory "from scratch" in purely operator theoretic terms would lengthen this treatise enormously and, while suitable for an expository text, would be inappropriate for a research memoir.

PROPOSITION 1.12. If S and T are operators either $S \lesssim T$ or $T \lesssim S$ or S and T admit direct sum decompositions, $S = S_1 \oplus S_2$ and $T = T_1 \oplus T_2$ such that $S_1 \overset{|}{\circ} S_2$ and $T_1 \overset{|}{\circ} T_2$ and $S_1 \lesssim T_1$ and $T_2 \lesssim S_2$.

DEFINITION 1.13. An operator T will be called a factor operator if it cannot be expressed as the direct sum of two disjoint suboperators.

Every irreducible operator is a factor operator but not conversely. Factor operators are the "minimal atoms" relative to disjoint direct sum decompositions in the way irreducible operators may be considered to be the "minimal atoms" in ordinary direct sum decompositions.

COROLLARY 1.14. Let T be a factor operator. If S is an operator such that $S \approx T$, then either $S \lesssim T$ or $T \lesssim S$.

DEFINITION 1.15. An operator T is infinite if $T \simeq \textcircled{\infty} T$. An operator T is finite if it contains no infinite suboperators.

These terms should not be confused with dimensional considerations. Infinite operators do have infinite dimensional domains, but not all operators on infinite dimensional spaces are infinite. There is even one (but only one) infinite operator that has finite rank, namely the zero operator on an infinite dimensional space. All operators acting on finite dimensional spaces are finite, but there exist finite operators acting on infinite dimensional spaces. (The unilateral shift is a finite operator, as is every irreducible operator.)

We apologize for using the terminology "finite operator," when it is inconsistent with the term "finite operator" recently introduced by James Williams [128] for an important class of operators which he has analyzed extensively. However for clarity and consistency with the von Neumann terminology it is crucial that we use the nomenclature of definition 1.15 (cf. proposition 1.38). Throughout this memoir there should be no confusion as we shall have no occasion to refer to the class of operators studied by Williams in

[128]. In any future work if confusion is likely to arise one may refer to "von Neumann-finite operators" as against "Williams-finite operators," or simply "Williams operators."

Every suboperator of a finite operator is finite. Every operator T is quasi-equivalent to one and, up to unitary equivalence, only one infinite operator, namely $\textcircled{\infty} T$ (cf. proposition 1.11).

> PROPOSITION 1.16. (Cf. Theorem 1.3 of [94]) If T is an operator, then either it is finite, or it is infinite or it admits a unique direct sum decomposition $T = T_1 \oplus T_2$ such that $T_1 \overset{|}{\circ} T_2$ and T_1 is finite and T_2 is infinite.

The Schröder-Bernstein result (theorem 1.3) indicates that the notion of unitary equivalence for operators is analogous to the notion of one-to-one correspondence for sets. We exploit this similarity further by giving a characterization of finite operators which bears a remarkable resemblance to the well-known characterization of finite sets.

> PROPOSITION 1.17. An operator T is finite if and only if it is not unitarily equivalent to any proper suboperator of itself.

PROOF. If T is not finite, then T admits an infinite suboperator and thus a suboperator of the form $\textcircled{\infty} T_0$. Hence T may be expressed in the form $T = T_1 \oplus \textcircled{\infty} T_0$. Clearly we may further decompose $\textcircled{\infty} T_0 = T_2 \oplus T_3$ where $T_2 \sim T_3 \simeq \textcircled{\infty} T_0$. Thus $T_1 \oplus T_2$ is a proper suboperator of T and $T_1 \oplus T_2 \simeq T_1 \oplus \textcircled{\infty} T_0 = T$.

Conversely suppose T is unitarily equivalent to a proper suboperator T_1 of T. Then T may be written in the form $T = S_1 \oplus T_1$ for some operator S_1. Since $T_1 \simeq T$, T_1 must also admit such a decomposition, i.e., $T_1 = S_2 \oplus T_2$ where $S_2 \simeq S_1$ and $T_2 \simeq T_1 \simeq T$. Proceeding inductively we may define a sequence of suboperators S_k of T defined on orthogonal subspaces of $\mathcal{H}(T)$ for which $S_k \simeq S_1$ for every k. Hence $\sum_{k=1}^{\infty} \oplus S_k$ is an infinite operator (being unitarily equivalent to $\textcircled{\infty} S_1$) which is a suboperator of T.

DEFINITION 1.18. Any operator quasi-equivalent to a finite operator will be called <u>semi-finite</u>.

An operator which has no finite suboperators will be called <u>purely infinite</u>.

> PROPOSITION 1.19. If T is an operator then either it is semi-finite or purely infinite or admits a unique direct sum decomposition $T = T_1 \oplus T_2$ such that $T_1 \mathrel{\circ\!\!|} T_2$ and T_1 is semi-finite and T_2 is purely infinite.

We have the following alternate characterization of purely infinite operators in terms of the equivalence relations introduced in the previous section.

> PROPOSITION 1.20. An operator T is purely infinite if and only if every operator quasi-equivalent to T is in fact unitarily equivalent to T.

PROOF. Suppose first that T is purely infinite and S is an operator quasi-equivalent to T. Since T has no finite suboperators, proposition 1.16 implies T is infinite, i.e., $T \simeq \textcircled{\infty} T$. By proposition 1.11 we have $\textcircled{\infty} S \simeq \textcircled{\infty} T \simeq T$. Clearly every suboperator of a purely infinite operator is also purely infinite and hence $S \lesssim T$ implies S is purely infinite and hence infinite. Thus $S \simeq \textcircled{\infty} S \simeq T$.

Conversely suppose that every operator quasi-equivalent to T is in fact unitarily equivalent to T. Then T is infinite since $\textcircled{\infty} T$ is quasi-equivalent to T and hence unitarily equivalent to T. If T is *not* purely infinite it has a finite suboperator say S. Let R denote the maximal suboperator of T disjoint from S. Then $S \oplus R$ is an operator quasi-equivalent to T. But $S \oplus R$ cannot be unitarily equivalent to T since $S \oplus R$ is not infinite.

DEFINITION 1.21. We say an operator T is *multiplicity free* if whenever two suboperators of T act on orthogonal subspaces, they are disjoint.

We say an operator is *homogeneous of order n* if $T \simeq \textcircled{n} M$ for a multiplicity free operator M. The letter n represents a positive integer or ∞.

DEFINITION 1.22. We say an operator is *discrete* if it is quasi-equivalent to a multiplicity free operator.

An operator T is said to be *of continuous type* if no suboperator of T is discrete.

> PROPOSITION 1.23. Every operator T is either discrete or of continuous type or has a unique direct sum decomposition $T = T_1 \oplus T_2$ such that $T_1 \overset{|}{\circ} T_2$, T_1 is discrete and T_2 is of continuous type.

Suboperators of continuous type operators behave remarkably like angleworms. One can always cut them in half and end up with two equivalent worms.

This characterization justifies the employment of the term "continuous" in this context.

PROPOSITION 1.24. An operator T is of continuous type if and only if every suboperator S of T may be decomposed in the form $S = S_1 \oplus S_2$ with $S_1 \simeq S_2$.

PROPOSITION 1.25. Every discrete operator is semi-finite and every purely infinite operator is continuous.

PROPOSITION 1.26. Two multiplicity free operators are quasi-equivalent if and only if they are unitarily equivalent.

PROPOSITION 1.27. Every discrete operator admits a unique decomposition of the form

$$T = \textcircled{\scriptsize\infty} M_0 \oplus \sum_{n=1}^{\infty} \oplus \textcircled{\scriptsize n} M_n$$

where $\{M_n\}$ is a set of mutually disjoint multiplicity free operators, $0 \leq n < +\infty$. Thus T has a unique disjoint direct sum decomposition into homogeneous operators of various orders. Of course all orders need not occur.

DEFINITION 1.28. (Type classification for operators) An operator is said to be <u>type I</u> if it is discrete. It is <u>type II</u> if it is semi-finite and of continuous type. A purely infinite operator is said to by <u>type III</u>.

If n is a positive integer or ∞, an operator is said to be <u>type I_n</u> if it is homogeneous of order n.

A finite operator of continuous type is said to be of <u>type II_1</u>.

An infinite operator of type II is said to be of <u>type II_∞</u>.

Clearly the classification into types I, II or III is invariant with respect to quasi-equivalence. Operators cannot be of more than one type. We warn the reader that the above terminology is not consistent with that of [117]. See corollary 1.40 below.

PROPOSITION 1.29. Every operator T has a unique decomposition of the form

$$T = T_{I} \oplus T_{II} \oplus T_{III}$$

where the operators T_i are mutually disjoint and T_i is of type i, for $i = I, II, III$. Of course some direct summands may not appear.

PROOF. First decompose $T = T_I \oplus S$ where T_I is discrete (i.e., type I), S is of continuous type and $T_I \mathrel{\rlap{\big|}\circ} S$, by proposition 1.23. Then decompose $S = T_{II} \oplus T_{III}$ where $T_{II} \mathrel{\rlap{\big|}\circ} T_{III}$, T_{II} is semi-finite and continuous and T_{III} is purely infinite by proposition 1.19. The uniqueness of the decomposition follows from the uniqueness of the decompositions given by propositions 1.19 and 1.23.

PROPOSITION 1.30. Every factor operator is of exactly one of the following types: I_n, $1 \le n \le \infty$, II_1, II_∞, III. Factor operators of each type exist.

DEFINITION 1.31. Let T be an operator. We define the <u>polynomial algebra of</u> T, denoted $\mathcal{P}(T)$, to be the $*$-algebra of all operators of the form $p(T,T^*)$ where $p(x,y)$ ranges over all complex polynomials in the two noncommuting variables x and y. The constant term a_o in $p(x,y)$ corresponds to the term a_oI, where I is the identity operator on the space $\mathcal{H}(T)$

of T, when one determines the operator $p(T,T^*)$.

DEFINITION 1.32. An operator T will be called <u>standard</u> if it is semi-finite and has a vector ψ in its domain $\mathcal{H}(T)$ such that

i) $\{S\psi : S \in \mathcal{R}(T)\}$ is dense in $\mathcal{H}(T)$

and

ii) $\{S\psi : S \in \mathcal{L}(\mathcal{H}(T)),\ ST = TS,\ ST^* = T^*S\}$ is dense in $\mathcal{H}(T)$.

PROPOSITION 1.33. Every semi-finite operator is quasi-equivalent to a standard operator. Two standard operators are quasi-equivalent if and only if they are unitarily equivalent. Thus every semi-finite quasi-equivalence class of operators contains, up to unitary equivalence, precisely one standard operator.

3. The bridge between operators and operator algebras

We next wish to make very explicit the basic connection of operator theory to von Neumann algebra theory so that we may cross back and forth with ease, throughout the rest of this memoir. There are two von Neumann algebras intimately associated with an operator T, the von Neumann algebra $\mathcal{A}(T)$ generated by T and the von Neumann algebra $\mathcal{A}(T)'$ of all operators on $\mathcal{H}(T)$ which commute with both T and T^*. Notice that we have obtained the entire classification theory for operators basically in terms of unitary equivalence, quasi-equivalence and the structure of the lattice of suboperators of the operator. Unitary equivalence and quasi-equivalence correspond to appropriate spatial and algebraic isomorphisms of the generated von Neumann algebras, while the lattice of suboperators corresponds to the projection lattice of the commuting algebra. Unitary equivalence of suboperators corresponds to the notion of equivalence of projections in the commuting algebra. The classification system described in the previous section then corresponds to the usual terminology for von Neumann algebras, applied sometimes to the generated algebra and other times to the commuting algebra. We proceed to make these statements explicit.

PROPOSITION 1.34. Two operators S and T are quasi-equivalent if and only if there exists a $*$-algebra isomorphism φ of $\mathcal{A}(S)$ onto $\mathcal{A}(T)$ such that $\varphi(S) = T$.

PROOF. (Cf. lemma 4 of [50]) Suppose $S \approx T$. By proposition 1.11 we have $\textcircled{\infty} S \simeq \textcircled{\infty} T$. Thus there exists a spatial isomorphism, call it φ_1, of $\mathcal{A}(\textcircled{\infty} S)$ onto $\mathcal{A}(\textcircled{\infty} T)$ such that $\varphi_1(\textcircled{\infty} S) = \textcircled{\infty} T$. Let φ_2 denote the natural $*$-isomorphism of $\mathcal{A}(S)$ onto $\mathcal{A}(\textcircled{\infty} S)$ defined by $\varphi_2(S) = \textcircled{\infty} S$ and similarly let φ_3 denote the natural $*$-isomorphism of $\mathcal{A}(T)$ onto $\mathcal{A}(\textcircled{\infty} T)$ defined by $\varphi_3(T) = \textcircled{\infty} T$. Then $\varphi = \varphi_3^{-1} \circ \varphi_1 \circ \varphi_2$ is a $*$-isomorphism of $\mathcal{A}(S)$ onto $\mathcal{A}(T)$ such that $\varphi(S) = T$.

Conversely suppose there exists a $*$-isomorphism φ of $\mathcal{A}(S)$ onto $\mathcal{A}(T)$ such that $\varphi(S) = T$. By the structure theorem (Theorem 3, page 55 of [37]) for $*$-algebra isomorphisms of von Neumann algebras $\varphi = \varphi_3 \circ \varphi_2 \circ \varphi_1$ where φ_1 is an amplification, $\mathcal{A}(S) \to \mathcal{A}(\textcircled{n} S)$ for some finite or countable cardinal n, φ_2 is an induction $\mathcal{A}(\textcircled{n} S) \to \mathcal{A}(\textcircled{n} S)_E$ where E is a projection in $\mathcal{A}(\textcircled{n} S)'$ and φ_3 is a spatial isomorphism of $\mathcal{A}(\textcircled{n} S)_E$ onto $\mathcal{A}(T)$. Thus T is unitarily equivalent to $\varphi_3^{-1}(T)$, which is of the form $\textcircled{n} S$ restricted to the range of E, a reducing subspace of $\textcircled{n} S$. Hence

$T \lesssim \textcircled{n} S$ and S covers T. By symmetry T also covers S, or $T \approx S$.

Recall (definition 1, page 215 of [37]) that two projections E and F of a von Neumann algebra $\mathcal{A}$ are said to be <u>equivalent relative to $\mathcal{A}$</u>, denoted $E \simeq F$ (re $\mathcal{A}$), if there exists an element U of $\mathcal{A}$ such that $U^*U = E$ and $UU^* = F$. (U is a partial isometry.) We write $E \lesssim F$ (re $\mathcal{A}$) if E is equivalent to a subprojection of F, relative to $\mathcal{A}$. The <u>central support</u> of a projection E in $\mathcal{A}$ is the smallest projection F in the center of $\mathcal{A}$ such that $E \leq F$.

If S is a suboperator of an operator T, where the domain spaces need not be separable, let $p(S)$ denote the projection of $\mathcal{H}(T)$ onto $\mathcal{H}(S)$. Then $S \to p(S)$ is a one-to-one order preserving map of the set of suboperators of T into the set of projections in $\mathcal{A}(T)'$. The map is almost surjective in the sense that every non-zero projection in $\mathcal{A}(T)'$ corresponds to a suboperator of T.

PROPOSITION 1.35. Let R and S denote suboperators of the operator T acting on spaces which are not necessarily separable.

1. $R \simeq S$ if and only if $p(R) \simeq p(S)$ relative to $\mathcal{A}(T)'$.
2. $R \lesssim S$ if and only if $p(R) \lesssim p(S)$ relative to $\mathcal{A}(T)'$.
3. If $T = R \oplus S$ then $R \overset{|}{\circ} S$ if and only if $p(R)$ is contained in the center of $\mathcal{A}(T)'$.
4. $R \approx S$ if and only if $p(R)$ and $p(S)$ have the same central support.
5. R covers S ($R \} S$) if and only if the central support of $p(R)$ is greater than or equal to the central support of $p(S)$.

PROOF. The first two parts are easy verifications and we first prove statement 3. Suppose that $R \underset{\circ}{|} S$. Let A be any operator in $\mathcal{A}(T)'$. Then A may be given a matricial representation

$$A = \begin{pmatrix} EAE & EA(I-E) \\ (I-E)AE & (I-E)A(I-E) \end{pmatrix}.$$

where $(I-E)AE$ intertwines R and S and $EA(I-E)$ intertwines S and R. But since $R \underset{\circ}{|} S$, corollary 1.8 implies $(I-E)AE = 0$ and $EA(I-E) = 0$. Hence $EA = EAE = AE$. Thus E is in the center of $\mathcal{A}(T)'$.

Conversely suppose E is in the center of $\mathcal{A}(T)$. Let A be any operator which intertwines R and S. Then we may consider A to be an operator on $\mathcal{H}(T)$ of the form $A = (I-E)AE$. Since A intertwines $R = TE$ and $S = T(I-E)$ we have

$$(I-E)AE(TE) = T(I-E)(I-E)AE$$

and

$$(I-E)AE(T^*E) = T^*(I-E)(I-E)AE$$

or

$$(I-E)AET = T(I-E)AE$$

and

$$(I-E)AET^* = T^*(I-E)AE.$$

Thus $(I-E)AE \in \mathcal{A}(T)'$. Since E is contained in the center of $\mathcal{A}(T)'$, E commutes with $(I-E)AE$. Thus $A = (I-E)AE = (I-E)AEE = E(I-E)AE = 0$. Thus every operator which intertwines R and S is necessarily zero. By corollary 1.8 we have $R \underset{\circ}{|} S$.

We next proceed to prove statement 4. Suppose first that R and S are quasi-equivalent and let F and G denote the central supports of $p(R)$ and $p(S)$ respectively. Since G is in the center, part 3 of this proposition implies that the restriction of T to the range of G is disjoint from the restriction of T to the range of $(I-G)$. Since S is a suboperator of the restriction of T to the range of G (since $p(S) \leq G$), S is also disjoint from the restriction of T to the range of $(I-G)$. Since $R \approx S$, R is also disjoint from the restriction of T to the range of $(I-G)$. Thus $p(R)$ must be orthogonal to $(I-G)$. Hence $p(R) \leq G$. Since $p(R) \leq F$ and F and G are central projections and hence commute we have that FG is a central projection for which $p(R) \leq FG$. Since F is the central support of $p(R)$ we have $F \leq FG$ and thus $F = FG$. By symmetry we have $G = FG$ and thus $F = G$.

Now suppose $p(R)$ and $p(S)$ have the same central support, say F. We will show that R covers S (and hence by symmetry that S covers R and thus $R \approx S$). Suppose R does not cover S. Then S admits a suboperator, say S_0, such that $R \overset{|}{\circ} S_0$. Let G denote the central support of $p(S_0)$. By lemma 1, page 217 of [37], applied to the projections $p(R)$ and $p(S_0)$, we have that F and G are orthogonal central projections. But $S_0 \leq S$ implies that $G \leq F$, since F is the central support of S. Hence $G = 0$ and thus $p(S_0) = 0$, which contradicts the fact that S_0 is a suboperator of S.

Statement 5 is proven in a manner similar to the proof of part 4.

COROLLARY 1.36. An operator T is a factor operator if and only if $\mathcal{A}(T)$ is a factor von Neumann algebra (i.e., if and only if the center of $\mathcal{A}(T)$ consists only of scalar multiples of the identity operator).

COROLLARY 1.37. An operator T is multiplicity free if and only if $\mathcal{A}(T)'$ is abelian.

In the context of von Neumann algebras the terms finite, properly infinite, semi-finite and purely infinite are given in definition 5, page 97 of [37]. The terms discrete (or type I) and continuous type are given in definition 1, page 121 of [37]. The notion of standard von Neumann algebra is given by definition 7, page 79 of [37]. Finally the notion of a type I_n (homogeneous of order n) von Neumann algebra is given by definition 1, page 239 of [37]. For the purposes of the following theorem, the terms infinite and properly infinite are synonymous.

PROPOSITION 1.38. An operator T is X if and only if its commuting algebra $\mathcal{A}(T)'$ is X, where X may denote any of the terms finite, infinite, semi-finite, purely infinite, discrete (or type I), of continuous type, standard, homogeneous of order n.

PROOF. According to theorem 1, page 288 of [37] a von Neumann algebra is finite if and only if every projection in the algebra, which is equivalent to the identity, is in fact equal to the identity. Applying this characterization to $\mathcal{A}(T)'$ and invoking propositions 1.35 and 1.17 we see that T is finite if and only if $\mathcal{A}(T)'$ is a finite von Neumann algebra.

A von Neumann algebra $\mathcal{A}$ is properly infinite if and only if the identity of $\mathcal{A}$ may be written as the sum of an infinite family of non-zero, two by two orthogonal, equivalent projections. This fact can be obtained by apply- proposition 10, page 235 of [37] and corollary 2, page 298 of [37] to the case where the projection E referred to in both statements is the identity of $\mathcal{A}$. We may now invoke proposition 1.35 to conclude that T is an infinite operator if and only if $\mathcal{A}(T)'$ is a properly infinite von Neumann algebra.

Suppose T is semi-finite. Then there exists a finite operator S such that T is quasi-equivalent to S. By the first paragraph of this proof we know that $\mathcal{A}(S)'$ is a finite von Neumann algebra. Further by proposition 1.34 there exists a $*$-algebra isomorphism φ of $\mathcal{A}(T)$ onto $\mathcal{A}(S)$ such that $\varphi(T) = S$. We may now apply corollary 3, page 233 of [37] to conclude that $\mathcal{A}(T)$ is a semi-finite von Neumann algebra. On the other hand if $\mathcal{A}(T)$ is semi-finite then corollary 3, page 233 of [37] states that there exists a von Neumann algebra, say $\mathcal{B}$ such that $\mathcal{B}'$ is finite and a $*$-algebra isomorphism φ of $\mathcal{A}(T)$ onto $\mathcal{B}$. Let $S = \varphi(T)$. Then $\mathcal{B} = \mathcal{A}(S)$, S is a finite operator (by the first paragraph of this proof) and T is quasi-equivalent to S by proposition 1.34. Thus we have established that T is a semi-finite operator if and only if $\mathcal{A}(T)$ is a semi-finite von Neumann algebra. But now corollary 1, page 104 of [37] implies that T is a semi-finite operator if and only if $\mathcal{A}(T)'$ is semi-finite.

The assertion that T is a purely infinite operator if and only if $\mathcal{A}(T)'$ is a purely infinite von Neumann algebra follows from proposition 11, page 235 of [37], applied to the von Neumann algebra $\mathcal{A}(T)'$ in the case where the projection mentioned there is the identity, and an invocation of proposition 1.35.

The fact that T is discrete (or type I) if and only if $\mathcal{A}(T)$ is discrete follows easily from proposition 1.34 and 1.37. The corresponding statement for $\mathcal{A}(T)'$ follows from theorem 1, page 123 of [37]. The fact that T is of continuous type if and only if $\mathcal{A}(T)$ is of continuous type now follows directly from proposition 1.35. The corresponding assertion for $\mathcal{A}(T)'$ then follows from corollary 1, page 124 of [37].

It follows from theorem 5, page 224 of [37] and corollary 1, page 6 of [37] that T is a standard operator if and only if $\mathcal{A}(T)$ is a standard von Neumann algebra. The corresponding assertion for $\mathcal{Q}(T)'$ is obtained from proposition 10 page 79 of [37].

Finally the assertion for homogeneity of order n simply follows from definitions and proposition 1.35.

COROLLARY 1.39. An operator T is X if and only if the von Neumann algebra $\mathcal{Q}(T)$ which it generates is X, where X may denote any one of the terms semi-finite, purely infinite, discrete, of continuous type, standard, type I, II or III.

PROOF. For each of the given terms, a von Neumann algebra has the property if and only if its commutator has the property (cf. [37], pages 79, 104, 123, 124, 126).

COROLLARY 1.40. An operator T is a factor operator of type I_n (respectively I_∞, II_1, II_∞, III) if and only if its commuting algebra $\mathcal{Q}(T)'$ is a factor von Neumann algebra of type I_n (respectively I_∞, II_1, II_∞, III).

Having made the connection between operator theory and von Neumann's theory of operator algebras explicit, we may now examine the earlier stated results from this point of view and supply the proofs which were deferred in the previous section. Thus theorem 1.3 may now be recognized as the counterpart of a statement about equivalence of projections within a von Neumann algebra (cf. [98] or proposition 1, page 216 of [37] or lemma 13 of [93]).

PROOF OF PROPOSITION 1.12. Let E' and F' denote the projections of $\mathcal{H}(S) \oplus \mathcal{H}(T)$ onto $\mathcal{H}(S)$ and $\mathcal{H}(T)$ respectively. Then E' and F' are both projections in the von Neumann algebra $\mathcal{A}(S) \oplus \mathcal{A}(T)$. We now apply theorem 1, page 218 of [37] to this von Neumann algebra to conclude that there exists a projection G in the center of $\mathcal{A}(S) \oplus \mathcal{A}(T)$ such that $F'G \precsim E'G$ (re $\mathcal{A}(S) \oplus \mathcal{A}(T)$) and $E'(I-G) \precsim F'(I-G)$ (re $\mathcal{A}(S) \oplus \mathcal{A}(T)$). Considering $\mathcal{A}(S)$ and $\mathcal{A}(T)$ to be embedded in $\mathcal{A}(S) \oplus \mathcal{A}(T)$ in the canonical way, we have that $E = GE'$ is in the center of $\mathcal{A}(S)$ and $F = GF'$ is in the center of $\mathcal{A}(T)$. Let S_1 (respectively S_2) denote the restriction of S to the range of $I-E$ (respectively E). Similarly let T_1 (respectively T_2) denote the restriction of T to the range of $I-F$ (respectively F). Then by proposition 1.35 we have $S_1 \overset{|}{\circ} S_2$, $T_1 \overset{|}{\circ} T_2$ and $F \precsim E$ (re $\mathcal{A}(S) \oplus \mathcal{A}(T)$) implies $T_2 \precsim S_2$ and $E'(I-G) \precsim F'(I-G)$ (re $\mathcal{A}(S) \oplus \mathcal{A}(T)$) implies $S_1 \precsim T_1$. We remark that a direct proof, purely in terms of the operator theoretic concepts of the first two sections of this chapter, could be modeled on the proof of theorem 1, page 218 of [37].

The remaining facts of section 2 may be obtained easily from the corresponding well-known facts for von Neumann algebras, using the bridge consisting of propositions 1.34, 1.35, 1.38 and their corollaries. The following table makes this correspondence explicit.

Proposition	Corresponding fact in [37]
1.16	proposition 8, page 98
1.19	proposition 8, page 98
1.23	corollary 1 to proposition 1, page 122
1.24	corollary 3, page 219 (cf. also lemma 4.12 of [87])
1.25	proposition 2 and its second corollary, page 122
1.26	corollary 1, page 241
1.27	proposition 2, page 240
1.33	corollary to proposition 9, page 100 and theorem 4, page 93

We conclude with a proof of proposition 1.30. The first assertion follows from corollary 1.40 and the well-known fact that every factor von Neumann algebra is exactly one of the types I_n, $1 \leq n \leq +\infty$, II_1, II_∞, III. We thus turn our attention to the problem of the existence of operators of each type. Fortunately a great deal of research has been done on the question of which von Neumann algebras are singly generated (cf. [43], [104], [107], [116], [117], [125], [130], [131], [132], [129]). The remarkable state of affairs is that all type I [104], and all properly infinite [130], and all hyperfinite [116] von Neumann algebras are singly generated. It remains an open question as to whether <u>every</u> von Neumann algebra is singly generated. It remains only to determine whether type II_1 factor von Neumann algebras, which are not hyperfinite, are single generated. These facts do not quite establish that factor operators of all types exist as the characterization of types (corollary 1.40) is in terms of the commuting algebra, rather than the generated algebra.

Notice that an operator is irreducible if and only if it is both multiplicity free and a factor. Thus if S is any irreducible operator then Ⓝ S is a factor operator of type I_n for $1 \leq n \leq +\infty$. (Further every type I_n operator is of this form.) Since every properly infinite von Neumann algebra is singly generated and type III (purely infinite) factor von Neumann algebras exist (and are properly infinite) it follows from corollary 1.39 that type III factor operators exist.

We know type II_∞ factor von Neumann algebras exist and ([130]) are singly generated. Thus there exists a type II operator by corollary 1.39. But by definition a type II operator is semi-finite and hence quasi-equivalent to a finite operator T. Thus T is a type II_1 operator and $\textcircled{\infty}$T is a type II_∞ operator. Thus operators of all types exist.

The existing literature does give some additional information. Normal operators are clearly type I. Warren Wogen [131], [132] has shown that the following classes of operators contain operators of each of the types I, II or III: hyponormal, nilpotent and transcendental quasi-nilpotent operators, unimodular contractions, operators similar to self-adjoint operators and operators similar to unitary operators. There exist partial isometries of all types [107], [116].

4. Multiplicity theory for factor operators

If T is an operator we have already used the notation $\textcircled{n}T$, where n is a positive integer or infinity, to denote a direct sum of n copies of T. Since there are various concrete ways of giving such a construction this concept is natural only up to unitary equivalence, in the sense that the unitary equivalence class of $\textcircled{n}T$ is uniquely determined by n and the unitary equivalence class of T.

Rather remarkably, this notion may be extended to give a natural meaning to $\textcircled{\lambda}T$, where λ is a positive real number, consistent with the above notation, whenever T is an operator of continuous type. For example, if T is an operator of continuous type, then proposition 1.24 states that T can be decomposed $T = S_1 \oplus S_2$ such that $S_1 \simeq S_2$. It is not trivial, but true, that the suboperator S_1 of T is determined by this property up to unitary equivalence. Cf. problem 1 of [86] or the corollary to lemma 44 of [93]. (Note that the lattice of suboperators of an operator T, being isomorphic to the lattice of projections in $\mathcal{Q}(T)'$, satisfies the axioms of Loomis' theory, by section 9 of [93].) Thus it is natural to denote (the unitary equivalence class of) S_1 by $\textcircled{\frac{1}{2}}T$. Continuing the division process according to proposition 1.24 we may define the operator $\textcircled{2^{-n}}$ T, up to unitary equivalence, for any integer n. Taking direct sums of such operators and employing the notation of the first paragraph of this section, we may define $\textcircled{r}$T, up to unitary equivalence, for any positive dyadic rational, i.e., any positive rational number whose denominator is of the form 2^n for some positive integer n. To circumvent the minor technicalities associated with showing $\textcircled{r}$T is well defined we can insist that r be reduced to lowest terms, i.e., that its numerator is odd.

PROPOSITION 1.41. Let T be an operator of continuous type and let λ be a positive real number. Then there exists an operator, which we shall denote $\textcircled{\lambda}$ T, which is determined uniquely, up to unitary equivalence, having the property that $\textcircled{r}$ T $\lesssim$ $\textcircled{\lambda}$ T for all positive dyadic rationals r less than λ and $\textcircled{\lambda}$ T $\lesssim$ $\textcircled{r}$ T for all dyadic rationals r greater than λ. If $\lambda \neq 1$, then $\textcircled{\lambda}$ T $\simeq$ T if and only if T is an infinite operator.

PROOF. (Cf. theorem 1.17 of [95]) Since the dyadic rationals are dense in the real line we may construct an increasing sequence $\{r_n\}$ of positive dyadic rationals which converges to λ. Then for each n, there exists a projection E_n in $\mathcal{A}(\textcircled{\infty}T)'$ such that the restriction of $\textcircled{\infty}$T to the range of E_n is unitarily equivalent to $\textcircled{r_n}$T, since $\textcircled{r_n}$T $\leq$ $\textcircled{\infty}$T. Proceeding inductively we can find projections F_n in $\mathcal{A}(\textcircled{\infty}T)'$ such that $F_n \simeq E_n$ in $\mathcal{A}(\textcircled{\infty}T)'$ and $\{F_n\}$ is an increasing sequence of projections in $\mathcal{A}(\textcircled{\infty}T)'$. Let F denote the least upper bound of the F_n in $\mathcal{A}(\textcircled{\infty}T)'$. Applying proposition 1.35 we conclude that the operator S which is the restriction of $\textcircled{$\infty$}T$ to the range of F has the property that $\textcircled{r}T \lesssim S$ for all dyadic rationals less than λ and $S \lesssim \textcircled{r}T$ for all dyadic rationals greater than λ. Since the projection F is determined up to equivalence, the operator S, which we henceforth denote $\textcircled{$\lambda$}T$, is determined up to unitary equivalence. Of course if T is infinite then, for dyadic rationals r_1 and r_2 with $r_1 < \lambda < r_2$ we have

$$\textcircled{∞}T \simeq \textcircled{r_1}T \lesssim \textcircled{λ}T \lesssim \textcircled{r_2}T \simeq \textcircled{∞}T$$

and hence that $\textcircled{$\lambda$}T \simeq \textcircled{$\infty$}T \simeq T$. The converse is a little stickier.

Suppose λ is a positive real number different than 1. Then there exists a dyadic rational, say $r = m \cdot 2^{-n}$ which is strictly between 1 and λ. We suppose $\lambda < r < 1$, the argument being very similar if $\lambda > 1$. Then $\textcircled{$\lambda$}T \lesssim \textcircled{r}T \lesssim T$ and hence $\textcircled{r}T \simeq T$. Thus $\textcircled{m}T \simeq \textcircled{2^n}T$. Since $m < 2^n$, $p = 2^n - m$ is a positive integer and

$$\textcircled{m}T \simeq \textcircled{m}T \oplus \textcircled{p}T .$$

Thus $\textcircled{m}T \simeq \textcircled{m}T \oplus \textcircled{kp}T$ for every positive integer k. From this fact we can construct a countable number of orthogonal subspaces of $\textcircled{m}T$ to establish $\textcircled{$\infty$}T \lesssim \textcircled{m}T$ and hence $\textcircled{m}T \simeq \textcircled{∞}T$. For any positive integer n we have

$$\textcircled{m+n}T \simeq \textcircled{m}T \oplus \textcircled{n}T \simeq \textcircled{∞}T \oplus \textcircled{n}T \simeq \textcircled{∞}T.$$

Thus for any pair of positive integers n, n' we have

$$\textcircled{m+n}T \simeq \textcircled{m+n$'$}T \simeq \textcircled{∞}T.$$

Thus for some k we have $\textcircled{$2^k$}T \simeq \textcircled{$2^{k+1}$}T \simeq \textcircled{$\infty$}T$. Thus again using (many times) problem 1 of [86], we have $T \simeq \textcircled{2}T$.

Continuing to double we see that $T \simeq \textcircled{2^k}T \simeq \textcircled{\infty}T$, i.e., T is infinite.

COROLLARY 1.42. Let T be an operator of continuous type and let λ and μ be two positive numbers such that $\lambda \neq \mu$. Then $\textcircled{\lambda}T \simeq \textcircled{\mu}T$ if and only if T is infinite.

Our next theorem describes the structure of the (unitary equivalence classes of the) suboperators of a factor operator (cf. definition 1.13). While we shall state and prove it in operator theoretic terms, the result corresponds to the dimension theory for projections in a factor von Neumann algebra (cf. proposition 14, page 236 of [37]). Note that by the first part of the theorem every normal factor operator is of type I_n for some n and is precisely of the form a scalar times the identity operator on an n-dimensional space.

THEOREM 1.43. If T is a factor operator of type I_n, $1 \leq n \leq +\infty$, then T has a irreducible suboperator T_o, and all the suboperators of T (up to unitary equivalence) are of the form $\textcircled{m}T_o$ for m a positive integer, $1 \leq m \leq n$. In particular $T \simeq \textcircled{n}T_o$. If T is a type II_1 factor operator, then all the suboperators of T (up to unitary equivalence) are of the form $\textcircled{\lambda}T$ where λ is a real number, $0 < \lambda \leq 1$. Thus the unitary equivalence classes of suboperators of T are parameterized in a one-to-one order preserving manner by the half open interval $(0,1]$.

If T is a type II_∞ factor operator, then T has a finite suboperator T_o and all the suboperators of T (up to unitary equivalence) are of the form $\textcircled{\lambda} T_o$, where $0 < \lambda \leq +\infty$, λ real. In particular $T \simeq \textcircled{\infty} T_o$. Thus the unitary equivalence classes of suboperators of T are parameterized in a one-to-one order preserving manner by the extended half line, $(0, +\infty]$.

If T is a type III factor operator, then every suboperator of T is unitarily equivalent to T.

PROOF. Suppose first that T is a factor of type I_n. Since it is homogeneous of order n, we have $T \simeq \textcircled{n} T_o$ for a multiplicity free operator T_o. Since T_o is quasi-equivalent to T by proposition 1.11, T_o is also a factor operator. By corollaries 1.36 and 1.37 $\mathcal{A}(T_o)'$ consists only of scalar multiples of the identity, i.e., T_o is irreducible.

If T is a type II_1 factor operator, then proposition 1.41 shows that $\textcircled{\lambda} T$, $0 < \lambda \leq 1$ are unitary equivalence classes of suboperators of T and if $0 < \lambda < \mu \leq 1$, then $\textcircled{\lambda} T \lesssim \textcircled{\mu} T$ and $\textcircled{\lambda} T$ and $\textcircled{\mu} T$ are not unitarily equivalent. Now let S be any suboperator of T. Since T is a factor operator, all suboperators of T are quasi-equivalent. Thus by corollary 1.14, for every λ, $0 < \lambda \leq 1$, either $S \lesssim \textcircled{\lambda} T$ or $\textcircled{\lambda} T \lesssim S$. Let

$$\mu = \operatorname{Sup}\{\lambda : 0 < \lambda \leq 1 \text{ and } \textcircled{\lambda} T \lesssim S\}.$$

Then proposition 1.41 implies $S \simeq \textcircled{\mu} T$.

If T is a type II_∞ factor operator, then it is an infinite operator which is quasi-equivalent to a finite operator S. By corollary 1.14, either

$S \lesssim T$ or $T \lesssim S$. Since T is infinite and S if finite the second inequality is impossible. Thus T contains a finite suboperator T_0 $(\simeq S)$. Proposition 1.41 shows that $\textcircled{\lambda} T_0$, $0 < \lambda \leq +\infty$, are unitary equivalence classes of suboperators of T and if $0 < \lambda < \mu \leq +\infty$, then $\textcircled{\lambda} T_0 \lesssim \textcircled{\mu} T_0$ and $\textcircled{\lambda} T_0$ and $\textcircled{\mu} T_0$ are not unitarily equivalent. Further since T_0 and T are quasi-equivalent we have $\textcircled{\infty} T_0 \simeq \textcircled{\infty} T \simeq T$ by proposition 1.11. Now let T_1 be an arbitrary suboperator of T. By corollary 1.14, for every λ, $0 < \lambda \leq +\infty$, either $T_1 \lesssim \textcircled{\lambda} T$ or $\textcircled{\lambda} T \lesssim T_1$. Let $\mu = \operatorname{Sup}\{\lambda : 0 < \lambda \leq +\infty$ and $\textcircled{\lambda} T \lesssim T_1\}$. Proposition 1.41 implies $T_1 \simeq \textcircled{\mu} T$.

The assertion for type III factor operators follows from a characterization of purely infinite (proposition 1.20) and the fact that every suboperator of a factor operator is quasi-equivalent to that operator.

Since the type classification (I, II and III) is an invariant of quasi-equivalence, we may interpret the multiplicity theory as describing the lattice of unitary equivalence classes of operators, within a quasi-equivalence class of factor operators

COROLLARY 1.44. A type I factor quasi-equivalence class contains precisely one (up to unitary equivalence) irreducible operator T_0 and all the unitary equivalence classes in the quasi-equivalence class are of the form $\textcircled{n} T_0$ where $n = \infty$ or $n = 1, 2, \ldots$.

A type II factor quasi-equivalence class always has a finite operator in it. If T_0 is any such finite operator, then all the unitary equivalence classes in the quasi-equivalence class are of the form $\textcircled{\lambda} T_0$ where λ is a positive real number or ∞.

A type III factor quasi-equivalence class consists of only one operator, up to unitary equivalence.

We remark that proposition 1.27 may be considered to be a multiplicity type statement for type I operators which are not factors. In general, multiplicity theory for operators which are not factor operators is considerably more complicated and will be taken up in chapter 4.

5. Weak containment and weak equivalence for operators

DEFINITION 1.45. If S is an operator, $C^*(S)$ will denote the C^*-algebra generated by S and the identity operator.

DEFINITION 1.46. If S and T are operators we say S is <u>weakly contained in</u> T, if there is a continuous $*$-homomorphism φ of $C^*(T)$ onto $C^*(S)$ such that $\varphi(T) = S$. We say two operators S and T are <u>weakly equivalent</u> (denoted $S \sim T$) if S is weakly contained in T and T is weakly contained in S.

It is immediate that the notion of weak equivalence of operators has a characterization analogous to the characterization for quasi-equivalence given in proposition 1.34. The two propositions together make it clear that any two quasi-equivalent operators are weakly equivalent.

PROPOSITION 1.47. Two operators S and T are weakly equivalent if and only if there exists a $*$-isomorphism φ of $C^*(S)$ onto $C^*(T)$ such that $\varphi(S) = T$.

The terminology "algebraic equivalence" of William Arveson [3] refers to the same notion as our weak equivalence while the terminology "algebraic equivalence" of Ping Kwan Tam [124] corresponds to what we have called quasi-equivalence. We have adopted the terminology of Michael Fell [56] which he has used in the context of representation theory. Our next result attempts to make the connections with representation theory explicit. If π_1 and π_2 are two $*$-representations of a C^*-algebra A, then π_1 is said to be weakly contained in π_2 if $\ker \pi_2 \subset \ker \pi_1$.

PROPOSITION 1.48. Let S and T denote two operators and let π_1 denote the $*$-representation of $C^*(T \oplus S)$ onto $C^*(T)$ such that $\pi_1(T \oplus S) = T$. Similarly let π_2 denote the $*$-representation of $C^*(T \oplus S)$ onto $C^*(S)$ such that $\pi_2(T \oplus S) = S$. We shall consider $\mathcal{H}(T)$ and $\mathcal{H}(S)$ to be embedded in $\mathcal{H}(T) \oplus \mathcal{H}(S)$ in the canonical way. Then the following are equivalent.

1) S is weakly contained in T

2) π_2 is weakly contained in π_1

3) π_1 is a faithful representation

4) Every positive linear functional on $C^*(T \oplus S)$ of the form $A \to (Ax,x)$ for a vector x in $\mathcal{H}(S)$ is a pointwise limit of finite linear combinations of positive linear functionals of the form $A \to (Ay,y)$ for vectors y in $\mathcal{H}(T)$.

5) Every positive linear functional on $C^*(T \oplus S)$ of the form $A \to (Ax,x)$ for a vector x in $\mathcal{H}(S)$ is a pointwise limit of finite sums of positive linear functionals of the form $A \to (Ay,y)$ for vectors y in $\mathcal{H}(T)$.

PROOF. We shall give a proof cycle showing that the first three conditions are equivalent. The equivalence of condition 2 to conditions 4 and 5 is then immediate from theorem 1.2 of [57] (cf. also theorem 3.4.4, page 66 of [36]).

Suppose that S is weakly contained in T. Let π denote the $*$-homomorphism of $C^*(T)$ onto $C^*(S)$ such that $\pi(T) = S$. Then $\pi_2 = \pi \circ \pi_1$ and thus $\ker \pi_1 \subset \ker \pi_2$, i.e., π_2 is weakly contained in π_1.

Next suppose π_2 is weakly contained in π_1, i.e., $\ker \pi_1 \subset \ker \pi_2$. Let E denote the projection of $\mathcal{H}(T) \oplus \mathcal{H}(S)$ onto $\mathcal{H}(T)$. Then

$$\ker \pi_2 = \{A : A \in C^*(T \oplus S) \text{ and } EA = AE = A\} \quad \text{and}$$
$$\ker \pi_1 = \{A : A \in C^*(T \oplus S) \text{ and } EA = AE = 0\}.$$

Hence $\ker \pi_1 \cap \ker \pi_2 = \{0\}$. Thus $\ker \pi_1 \subset \ker \pi_2$ implies $\ker \pi_1 = \{0\}$ or π_1 is faithful.

Finally if π_1 is a faithful and hence invertible $*$-homomorphism we have $\pi = \pi_2 \circ \pi_1^{-1}$ is a $*$-homomorphism of $C^*(T)$ onto $C^*(S)$ such that $\pi(T) = S$.

In the next chapter we shall introduce a new notion of the "spectrum" of an operator T, roughly speaking as the set of (unitary equivalence classes of) irreducible operators weakly contained in T. For that reason it is useful to have conditions under which an irreducible operator is weakly contained in another irreducible operator.

PROPOSITION 1.49. Let S and T denote irreducible operators. Consider the spaces $\mathcal{H}(T)$ and $\mathcal{H}(S)$ to be canonically embedded in $\mathcal{H}(T) \oplus \mathcal{H}(S)$. The following conditions are equivalent.

1. S is weakly contained in T.
2. For some nonzero vector x in $\mathcal{H}(S)$, the positive

linear functional $A \to (Ax,x)$, defined on $C^*(T \oplus S)$, is a pointwise limit of positive linear functionals on $C^*(T \oplus S)$ of the form $A \to (Ay,y)$ where $y \in \mathcal{H}(T)$.

3. For every unit vector x in $\mathcal{H}(S)$, there exists a net $\{y_\lambda\}$ of unit vectors in $\mathcal{H}(T)$ such that

$$(Ax,x) = \lim_\lambda (Ay_\lambda, y_\lambda)$$

for every element A of $C^*(T \oplus S)$.

PROOF. If we use the equivalence of conditions 1 and 2 of proposition 1.48 and apply known results for representations, the equivalences are immediate. Indeed this is just the equivalences of parts ii), iii), and iv) of theorem 3.4.10, page 68 of [36]. Note that in this case both of the representations described in proposition 1.48 are irreducible. (Compare also theorem 1.4 of [57].)

6. Smooth operators

DEFINITION 1.50. An operator T will be called smooth if all the operators weakly contained in T are type I.

PROPOSITION 1.51. An operator T is smooth if and only if $C^*(T)$ is GCR in the terminology of Irving Kaplansky [88], or "postliminaire" in the terminology of Jacques Dixmier [36] or type I in the terminology of James Glimm [69].

PROOF. Simply note that to every operator S weakly contained in T there corresponds a representation π_S of $C^*(T)$ for which $\pi_S(T) = S$ and to each $*$-representation π of $C^*(T)$ there corresponds an operator $\pi(T)$

weakly contained in T. Further a $*$-representation π is type I if and only if the algebra of operators commuting with the range of π is a type I von Neumann algebra. Thus (cf. corollary 1.40) T is smooth if and only if all the $*$-representations of $C^*(T)$ are type I. Since a singly generated C^*-algebra is separable we may apply theorem 9.1 (i) $\Leftrightarrow$ iii)) page 168 of [36] to obtain the desired result.

Using this characterization some authors (for example [21] and [102]) have called such operators GCR operators. However all of the terms GCR (who can remember what such initials stand for?), "postliminaire" (how does one translate that?) or type I (we have already defined type I and there exist type I operators which are not smooth) suggested by proposition 1.51 seem undesirable. As we shall have ample opportunity to observe throughout this memoir the line between smooth and nonsmooth represents a sharp demarcation with smooth sailing on one side and rough waters on the other. However the justification for this choice of terminology derives from a characterization (theorem 2.32) to be obtained in the next chapter, of smooth operators as those whose (generalized) spectra satisfy a "smoothness" condition. We mention that the notion of smooth operator has nothing whatsoever to do with the notion of smoothing operators used in [99] or the notion of smooth integral operators used in [4].

The subsequent theory, particularly the spectral theory, decomposition theory and theory of spectral multiplicity, all take significantly simpler forms when restricted to the class of smooth operators. However we prefer to develop the general theory which applies to all operators, and then to indicate the simplifications which occur when an operator is known to be smooth (section 1 of chapter 5).

The following observation is immediate, but nevertheless useful.

PROPOSITION 1.52. Every operator weakly contained in a smooth operator is smooth.

PROPOSITION 1.53. There exists an irreducible operator (and hence type I) which is not smooth.

PROOF. By corollary 1 and its proof of [125] there exists an irreducible operator (henceforth called the Topping operator) which generates a uniformly hyperfinite C^*-algebra, which is known not to be GCR. For a concrete representation of the Topping operator see the remark following corollary 8 of [23].

While type I operators need not be smooth, Takateru Okayasu has shown that every type I von Neumann algebra on a separable Hilbert space is singly generated by a smooth operator [102]. Further he has shown that if T and S are smooth operators for which $TS = ST$ and $T^*S = ST^*$, then TS is also smooth. It is easy to see that T is smooth if and only if T^* is smooth. All isometries (and hence the unilateral shift) are smooth [121]. Further normal operators and compact operators are known to be smooth. Condition 3 of the next proposition shows that all n-normal operators are smooth and corollary 5.8 of this memoir shows all quasi-normal operators are smooth. This observation for n-normal operators is made in [21].

The following characterization of smooth operators corresponds basically to a famous theorem of James Glimm [69] (see also [40], [41], [56], and [88]).

PROPOSITION 1.54. Let A be an operator. Then the following conditions are equivalent:

1. A is smooth.

2. Every factor operator weakly contained in A is type I.

3. If S is an irreducible operator weakly contained in A, then $C^*(S)$ contains the compact operators on $\mathcal{H}(S)$.

4. If S and T are any two weakly equivalent irreducible operators weakly contained in A, then S and T are unitarily equivalent.

PROOF. We may argue exactly as we did to prove proposition 1.51 and apply theorem 9.1, page 168 of [36] to the C^*-algebra $C^*(A)$. One only need note, in condition 4 above, that S and T are weakly equivalent if and only if the corresponding representations of $C^*(A)$ have the same kernel.

COROLLARY 1.55. There exist irreducible operators which are weakly equivalent but not unitarily equivalent.

PROOF. This follows from the equivalence of 1 and 4 in the above proposition and the fact that there exist non-smooth operators.

COROLLARY 1.56. If S and T are smooth irreducible operators which are weakly equivalent, then they are unitarily equivalent.

PROOF. Since T and S are weakly equivalent proposition 1.48 ($1 \Leftrightarrow 3$) implies that the $*$-homomorphism $C^*(T \oplus S) \to C^*(T)$ is faithful. Thus T is weakly equivalent to $T \oplus S$ (proposition 1.47) and hence $T \oplus S$ is also smooth. We may now apply part 4 of the previous proposition to the operator $T \oplus S$.

7. The envelope of an operator

DEFINITION 1.57. Let T be an operator defined on a (not necessarily separable) Hilbert space $\mathcal{H}(T)$. The envelope of T, denoted $e(T)$, is the direct sum of all operators acting on some fixed separable space (say ℓ_2) which are weakly contained in T. The von Neumann algebra $\mathcal{A}(e(T))$ generated by $e(T)$ will be called the enveloping von Neumann algebra of T, and will be denoted $\mathcal{E}(T)$.

Clearly the operator $e(T)$ will in general act on a nonseparable space, even when $\mathcal{H}(T)$ is separable. Further $e(T)$ has the property that every operator on a separable space which is weakly contained in T, is unitarily equivalent to a suboperator of $e(T)$. There is a canonical $*$-homomorphism π of $C^*(T)$ into $\mathcal{E}(T)$ such that $\pi(T) = e(T)$. Thus $e(T)$ is weakly contained in T. Further T is the direct sum of (cyclic) operators weakly contained in $e(T)$ and hence is itself weakly contained in $e(T)$. Thus T is weakly equivalent to $e(T)$ and (proposition 1.47) $C^*(T)$ and $C^*(e(T))$ are isomorphic C^*-algebras.

Since the operators which are weakly contained in T and which act on a separable Hilbert space correspond precisely to the separable $*$-representations π of $C^*(T)$ by the correspondence $\pi \leftrightarrow \pi(T)$, and the direct sum of all such separable $*$-representations is quasi-equivalent to what Jacques Dixmier calls the universal representation of $C^*(T)$ (cf. section 2.7.6, page 43 of [36]) we have (using proposition 1.34) that $\mathcal{E}(T) = \mathcal{A}(e(T))$ is isomorphic to the weak closure of the range of the universal representation of $C^*(T)$, i.e., $\mathcal{E}(T)$ is isomorphic to the enveloping von Neumann algebra of the C^*-algebra $C^*(T)$. Hence all of the interesting material of paragraph

XII of [36] becomes applicable in this context (see also [51] and [52]). Clearly this construct is also closely related to the Berberian representation of an operator [11].

The construct, the envelope of an operator T, is useful in describing the set $Q(T)$ of all quasi-equivalence classes of all operators weakly contained in T, partially ordered by the covering relation. In describing the ordering of $Q(T)$ it is convenient to adjoin an element, denoted δ_o, to $Q(T)$ and to extend the partial ordering to this enlarged set, still denoted by $Q(T)$, by saying that every element of $Q(T)$ covers δ_o. (For example δ_o could be thought of as the quasi-equivalence class of the trivial operator on a zero dimensional space.) Thus two points α and β of $Q(T)$ are disjoint if and only if the infimum of α and β is equal to δ_o. This device will ensure that any two elements of $Q(T)$ will have an infimum and thus that $Q(T)$ will be a lattice. The following proposition corresponds to theorem 1 of [53].

PROPOSITION 1.58. If T is an operator, then the lattice $Q(T)$ of quasi-equivalence classes of operators weakly contained in T is lattice isomorphic to the lattice of projections in an abelian von Neumann algebra, specifically the center of the enveloping von Neumann algebra $\mathcal{E}(T)$ of T. (The space $\mathcal{H}(T)$ of T need not be separable.)

PROOF. Every operator weakly contained in T is unitarily equivalent to a direct sum of separable (cyclic) suboperators of $e(T)$ and thus corresponds (by taking a least upper bound) to a projection in $\mathcal{Q}(e(T))'$ by the correspondence of proposition 1.35. By part 4 of that proposition, quasi-equivalence classes of such operators correspond precisely to the central projections of $\mathcal{A}(e(T))'$. By part 5 of proposition 1.35 this one-to-one correspondence is order preserving.

PROPOSITION 1.59. The lattice $\tilde{\mathcal{L}}_1$ of quasi-equivalence classes of operators of norm less than or equal to 1, acting on a separable Hilbert space, and partially ordered with respect to the covering relation, is lattice isomorphic to a lattice of projections contained in an abelian von Neumann algebra. Thus $\tilde{\mathcal{L}}_1$ is a distributive σ-complete lattice which is closed under relative complements.

PROOF. Let T denote the direct sum of all operators of norm less than or equal to 1, acting on some fixed separable Hilbert space, say ℓ_2. Then we may apply proposition 1.57 to the operator T to conclude that $\tilde{\mathcal{L}}_1$ is in one-to-one correspondence with those projections in the center of $\mathcal{E}(T)$ which come from quasi-equivalence classes which contain operators acting on separable spaces.

Proposition 1.59 suggests that we might be able to characterize operators up to quasi-equivalence (i.e., $\tilde{\mathcal{L}}_1$) in terms of some lattice of finite measure classes, ordered by absolute continuity, on some appropriate measure space. It is for just such a purpose that we now turn our attention in chapter 2 to various technical matters relating to Borel structures for operators (cf. sections 47 and 48 of [77]).

Chapter 2

THE SPECTRUM AND QUASI-SPECTRUM OF AN OPERATOR

1. The $*$-strong operator topology and its associated Borel structure

Charles Akemann, in his thesis [1], has studied extensively the $*$-strong operator topology which we define and describe next. This operator topology, which we shall use throughout this memoir, is very natural and well behaved. Indeed Akemann has proved (theorem II.7 of [1]) that on the unit ball, the $*$-strong operator topology coincides precisely with the Mackey topology. This topology is also considered in [122].

DEFINITION 2.1. Let $\mathcal{L}(\mathcal{H})$ denote the algebra of all bounded linear operators on the Hilbert space $\mathcal{H}$. The *$*$-strong topology* on $\mathcal{L}(\mathcal{H})$ is the locally convex topology induced by the family of all semi-norms of the form $T \to \|T\psi\|$ or $T \to \|T^*\psi\|$, where $\psi \in \mathcal{H}$.

This strengthening of the ordinary strong operator topology leads to an operator topology which meshes with the $*$-algebra operations a bit better than either the weak or strong operator topology separately. Specifically the adjoint operation is (obviously) continuous while we also have that bounded multiplication is jointly continuous.

PROPOSITION 2.2. The adjoint map in $\mathcal{L}(\mathcal{H})$ is $*$-strong continuous. Further if, for $r > 0$, $\mathcal{L}_r(\mathcal{H})$ denotes the ball in $\mathcal{L}(\mathcal{H})$ of radius r, then the map $(S,T) \to ST$ of $\mathcal{L}_r(\mathcal{H}) \times \mathcal{L}_r(\mathcal{H})$ into $\mathcal{L}(\mathcal{H})$ is $*$-strong continuous. Further if $\mathcal{H}$ is separable, the ball $\mathcal{L}_r(\mathcal{H})$ of radius r is a complete and metrizable uniform space in the $*$-strong operator topology.

PROOF. Let $\{S_\lambda\}$ and $\{T_\lambda\}$ denote nets in $\mathcal{L}_r(\mathcal{H})$ converging to S_o and T_o respectively, in the $*$-strong operator topology. Thus $S_\lambda T_\lambda$ converges strongly to $S_o T_o$ by properties of the strong operator topology, i.e., $\|(S_\lambda T_\lambda - S_o T_o)\psi\| \to 0$ for every vector $\psi \in \mathcal{H}$. A similar argument goes as follows

$$\begin{aligned}\|(S_\lambda T_\lambda - S_o T_o)^*\psi\| &= \|(T_\lambda^* S_\lambda^* - T_o^* S_o^*)\psi\| \\ &\leq \|(T_\lambda^*(S_\lambda^* - S_o^*) + (T_\lambda^* - T_o^*)S_o^*)\psi\| \\ &\leq r\|(S_\lambda^* - S_o^*)\psi\| + \|(T_\lambda^* - T_o^*)(S_o^*\psi)\| \to 0.\end{aligned}$$

Thus bounded multiplication is $*$-strong jointly continuous.

Next let $\{\psi_i\}$ denote a countable set of vectors dense in $\mathcal{H}$. Then the countable family of pseudo-metrics $p_{1i}(S,T) = \|(T-S)\psi_i\|$ and $p_{2i}(S,T) = \|(T-S)^*\psi_i\|$, $i = 1,2,\ldots$, defines the $*$-strong uniformity on $\mathcal{L}_r(\mathcal{H})$. Indeed suppose $S_\lambda \to S$ relative to this countable family of pseudo-metrics and let ψ be an arbitrary element of $\mathcal{H}$. If $\varepsilon > 0$, choose ψ_i such that $\|\psi_i - \psi\| < \varepsilon/4r$ and choose N such that $\|(S_\lambda - S)\psi_i\| < \varepsilon/2$ and $\|(S_\lambda - S)^*\psi_i\| < \varepsilon/2$ whenever $\lambda > N$. Then

$$\begin{aligned}\|(S_\lambda - S)\psi\| &= \|(S_\lambda - S)(\psi_i + (\psi - \psi_i))\| \\ &\leq \|(S_\lambda - S)\psi_i\| + \|(S_\lambda - S)\|\,\|\psi - \psi_i\| \\ &< \frac{\varepsilon}{2} + 2r\left(\frac{\varepsilon}{4r}\right) = \varepsilon \quad \text{whenever } \lambda > N.\end{aligned}$$

Similarly $\|(S_\lambda - S)^*\psi\| < \varepsilon$ whenever $\lambda > N$. This shows that the countable family of pseudo-metrics defines the same $*$-strong operator topology.

Thus the uniformity on $\mathcal{L}_r(\mathcal{H})$ has a countable base and is clearly Hausdorff (being stronger than the strong operator topology which is already Hausdorff). Thus the $*$-strong operator topology on $\mathcal{L}_r(\mathcal{H})$ is metrizable (cf. page 186 of [91]).

Finally we verify that $\mathcal{L}_r(\mathcal{H})$ is complete in the $*$-strong uniformity. Suppose $\{T_\lambda\}$ is a Cauchy net in $\mathcal{L}_r(\mathcal{H})$. Recall (lemma 20, page 191 of [91]) that being Cauchy in a uniform space is equivalent to being Cauchy relative to each member of the family of pseudo-metrics defining the uniformity. Thus for each ψ in $\mathcal{H}$, $\{T_\lambda\psi\}$ and $\{T_\lambda^*\psi\}$ are Cauchy nets in $\mathcal{H}$, relative to the Hilbert space norm. Thus for each ψ in $\mathcal{H}$ there exists vectors φ_ψ and ξ_ψ such that $T_\lambda\psi \to \varphi_\psi$ and $T_\lambda^*\psi \to \xi_\psi$ strongly. Define an operator T in $\mathcal{L}(\mathcal{H})$ by $T\psi = \varphi_\psi$ for ψ in $\mathcal{H}$. Now it is easy to verify that T is a well-defined linear map on $\mathcal{H}$ and $\|T\| \leq r$. We conclude by verifying that $T_\lambda \to T$ in the $*$-strong operator topology. For each $\psi \in \mathcal{H}$ we have $\|(T_\lambda - T)\psi\| \to 0$ by the definition of T, and it thus remains to show that $\|(T_\lambda - T)^*\psi\| \to 0$. Since $\{T_\lambda^*\psi\}$ converges in the norm to ξ_ψ, it also converges weakly to ξ_ψ. Since we already have that $T_\lambda \to T$ in the strong operator topology, it also converges weakly and thus $T_\lambda^* \to T^*$ weakly and hence $T_\lambda^*\psi \to T^*\psi$ weakly. Since the weak topology in $\mathcal{H}$ is Hausdorff the limit must be unique, i.e., $T^*\psi = \xi_\psi$. Thus $\|T_\lambda^*\psi - T^*\psi\| \to 0$ and we are done.

Following the terminology of George Mackey [78] a σ-field of subsets of a set S will be called a Borel structure for S and the elements of the σ-field will be called the Borel sets of the Borel space S. The Borel structure generated by a topology will be the smallest Borel structure containing the closed sets. A function between two such Borel spaces is called a Borel function if the inverse images of Borel sets are Borel. A Borel space is said to be standard if it is Borel isomorphic to the Borel space associated with a Borel subset of a complete separable metric space. Two standard Borel spaces are Borel isomorphic if and only if they have the same cardinal number and the only infinite cardinals possible are $\aleph_0$ and $2^{\aleph_0}$. For the basic facts on Borel spaces see [96].

PROPOSITION 2.3. Let $\mathcal{H}$ be a separable Hilbert space and let $\mathcal{L}(\mathcal{H})$ denote the algebra of all bounded linear operators on $\mathcal{H}$. Then the three operator topologies, weak, strong and $*$-strong all generate the same Borel structure on $\mathcal{L}(\mathcal{H})$, which makes $\mathcal{L}(\mathcal{H})$ into a standard Borel space.

PROOF. By proposition 2.2, for each $r > 0$, the ball $\mathcal{L}_r(\mathcal{H})$ of radius r is complete and metrizable with respect to each of the three operator topologies. (Compare also chapter 1, §3 of [37].) Thus $\mathcal{L}_r(\mathcal{H})$ is a standard Borel space for each $r > 0$. Further, for each $r > 1$, $\mathcal{L}_{r-1}(\mathcal{H})$ is a Borel (in fact closed) subset of $\mathcal{L}_r(\mathcal{H})$, for any of the three topologies. Let $\mathcal{B}_1 = \mathcal{L}_1(\mathcal{H})$ and $\mathcal{B}_r = \{T : T \in \mathcal{L}_r(\mathcal{H}),\ T \notin \mathcal{L}_{r-1}(\mathcal{H})\}$ for $r = 2,3,\ldots$. Then $\{\mathcal{B}_r\}$ is a sequence of mutually disjoint standard Borel spaces and $\mathcal{L}(\mathcal{H}) = \bigcup_r \mathcal{B}_r$. By theorem 3.1 of [96] $\mathcal{L}(\mathcal{H})$ is a standard Borel space, regardless of which of the three topologies generated the Borel structure. Hence it follows from theorem 3.2 of [96] that the standard Borel structure on $\mathcal{L}(\mathcal{H})$ is the same, regardless of which operator topology was used to generate it.

We next note that this Borel structure in $\mathcal{L}(\mathcal{H})$ is quite compatible with the algebraic structure.

PROPOSITION 2.4. Addition, multiplication and the $*$-operation are all Borel operations in $\mathcal{L}(\mathcal{H})$. Further if $|T|$ denotes the absolute value $(T^*T)^{\frac{1}{2}}$ of an operator and E_T and F_T denote the projections onto the closure of the range of T and the orthogonal complement of the kernel of T respectively, then $T \to |T|$, $T \to E_T$ and $T \to F_T$ are all Borel maps of $\mathcal{L}(\mathcal{H})$ into $\mathcal{L}(\mathcal{H})$.

PROOF. (Cf. also page 1163 of [46]) The linear and $*$ - operations are $*$ - strong continuous and hence Borel. Further let m denote the multiplication map of $\mathcal{L}(\mathcal{H}) \times \mathcal{L}(\mathcal{H})$ into $\mathcal{L}(\mathcal{H})$. Let B denote a Borel subset of $\mathcal{L}(\mathcal{H})$. For each positive integer n,

$$m^{-1}(B) \cap \left(\mathcal{L}_n(\mathcal{H}) \times \mathcal{L}_n(\mathcal{H})\right)$$

is a Borel subset of $\mathcal{L}(\mathcal{H}) \times \mathcal{L}(\mathcal{H})$ since multiplication is $*$ - strong continuous on bounded sets. Hence

$$m^{-1}(B) = \bigcup_n \left\{m^{-1}(B) \cap \left(\mathcal{L}_n(\mathcal{H}) \times \mathcal{L}_n(\mathcal{H})\right)\right\}$$

is Borel.

Suppose f is a bounded real valued Borel function defined on the real line and let p_n denote a sequence of real polynomials converging to f point-wise, and which are uniformly bounded on compact sets. If A is a self-adjoint element (i.e., T^*T) of $\mathcal{L}(\mathcal{H})$, we have from the spectral theory that $p_n(A) \to f(A)$ weakly. Thus $A \to f(A)$ is Borel. Hence $T \to T^*T \to (T^*T)^{\frac{1}{2}} = |T|$ is Borel.

Let g be the characteristic function of the open set $(0, \infty)$. Then we have $T \to g((T^*T)^{\frac{1}{2}}) = F_T$ is Borel. Finally $T \to T^* \to F_{T^*} = E_T$ is Borel.

While it may appear to be a very technical matter it will nevertheless be useful to establish that certain important subsets of $\mathcal{L}(\mathcal{H})$ are in fact Borel sets (and hence in particular, by corollary 1, page 139 of [96] standard

Borel spaces). While it will continue to be a worthwhile and useful research program to establish that certain classes of operators form Borel sets in $\mathcal{L}(\mathcal{H})$, we shall in this section content ourselves with establishing that the following basic sets are indeed Borel sets: normal operators, unitary operators, irreducible operators, factor operators, type I operators, finite operators and factor operators of each of the types I_∞, II_1, II_∞, III. We shall first give a concrete computational proof that the set $\mathfrak{F}$ of all factor operators on a separable Hilbert space is in fact a $\mathfrak{F}_{\sigma\delta\sigma}$ set in the $*$-strong topology, the argument being essentially the same as used by Jacques Dixmier in a representation theoretic context (cf. theorem 7.1.4, page 135 of [36]).

LEMMA 2.5. Let d be a metric on the unit ball $\mathcal{L}_1(\mathcal{H})$ of bounded linear operators on the separable Hilbert space $\mathcal{H}$, compatible with the weak operator topology. Let S be an element of $\mathcal{L}_1(\mathcal{H})$ and let a, b, c denote non-negative real numbers. Let $p_1,\ldots,p_n$ denote complex polynomials in two noncommuting variables (cf. definition 1.31). Let Y denote the set of all operators T in $\mathcal{L}_c(\mathcal{H})$, the ball of radius c, having the property that there exist operators $R_1,\ldots,R_n$ in $\mathcal{L}_b(\mathcal{H}) \cap \mathcal{A}(T)'$ such that

$$\sum_{i=1}^{n} p_i(T,T^*)R_i \in \mathcal{L}_1(\mathcal{H})$$

and

$$d\left(\sum_{i=1}^{n} p_i(T,T^*)R_i\,,\, S\right) \leq a.$$

Then Y is $*$-strong closed in $\mathcal{L}(\mathcal{H})$.

PROOF. Let T^λ denote a net in Y, $\lambda \in \Lambda$ which converges to an operator T in the $*$-strong operator topology. Since $\|T^\lambda\| \leq c$ for all λ, we have $\|T\| \leq c$. Since $T^\lambda \in Y$, we have, for each λ in Λ, that there exist operators $R_1^\lambda, \ldots, R_n^\lambda$ in $\mathcal{L}_b(\mathcal{H}) \cap \mathcal{A}(T)'$ such that

$$\sum_{i=1}^n p_i(T^\lambda, T^{\lambda *})R_i^\lambda \in \mathcal{L}_1(\mathcal{H})$$

and

$$d\Big(\sum_{i=1}^n p_i(T^\lambda, T^{\lambda *})R_i^\lambda, S\Big) \leq a.$$

Since $\mathcal{L}_b(\mathcal{H})$ is weakly compact, one may, by taking a subnet of T^λ, assume that $R_1^\lambda, R_2^\lambda, \ldots, R_n^\lambda$ converge weakly to some elements $R_1, R_2, \ldots, R_n$ in $\mathcal{L}_b(\mathcal{H})$. Since $R_i^\lambda \in \mathcal{A}(T)'$, which is weakly closed, we have $R_i \in \mathcal{L}_b(\mathcal{H}) \cap \mathcal{A}(T)'$ for $i = 1, \ldots, n$. We next verify that $\sum_{i=1}^n p_i(T^\lambda, T^{\lambda *})R_i^\lambda$ converges weakly to $\sum_{i=1}^n p_i(T, T^*)R_i$. Indeed let $\psi, \varphi \in \mathcal{H}$. Then

$$\Big|\Big(\Big(\sum_{i=1}^n p_i(T^\lambda, T^{\lambda *})R_i^\lambda - \sum_{i=1}^n p_i(T, T^*)R_i\Big)\psi, \varphi\Big)\Big|$$

$$\leq \Big|\Big(\Big(\sum_{i=1}^n p_i(T^\lambda, T^{\lambda *})R_i^\lambda - \sum_{i=1}^n p_i(T, T^*)R_i^\lambda\Big)\psi, \varphi\Big)\Big|$$

$$+ \Big|\Big(\Big(\sum_{i=1}^n p_i(T, T^*)R_i^\lambda - \sum_{i=1}^n p_i(T, T^*)R_i\Big)\psi, \varphi\Big)\Big|$$

$$\leq \Big\|\sum_{i=1}^n \Big(p_i(T^\lambda, T^{\lambda *}) - p_i(T, T^*)\Big)\psi\Big\| \cdot \|R_i^\lambda \varphi\|$$

$$+ \Big|\sum_{i=1}^n \Big((R_i^\lambda - R_i)\psi, (p_i(T, T^*))^* \varphi\Big)\Big|$$

$$\leq b\|\varphi\| \sum_{i=1}^n \Big\|\Big(p_i(T^\lambda, T^{\lambda *}) - p_i(T, T^*)\Big)\psi\Big\|$$

$$+ \sum_{i=1}^n \Big|\Big((R_i^\lambda - R_i)\psi, (p_i(T, T^*))^* \varphi\Big)\Big| .$$

Now the second of these two summands converges to zero since R_i^λ converges to R_i weakly, for $i=1,\ldots,n$. The first of the summands converges to zero since the basic algebraic operations are $*$-strongly continuous on bounded sets (proposition 2.2), $\|T^\lambda\| \leq c$ for all λ and hence $p_i(T^\lambda, T^{\lambda*})$ converges $*$-strongly and hence strongly to $p_i(T,T^*)$ for $i=1,\ldots,n$.

Now since each $\sum_{i=1}^n p_i(T^\lambda, T^{\lambda*})R_i^\lambda$ is in the unit ball $\mathcal{L}_1(\mathcal{H})$ we also have that $\sum_{i=1}^n p_i(T,T^*)R_i \in \mathcal{L}_1(\mathcal{H})$. Since the metric d is compatible with the weak operator topology and $d(\sum_{i=1}^n p_i(T^\lambda, T^{\lambda*})R_i^\lambda, S) \leq a$ for every λ, we have that $d(\sum_{i=1}^n p_i(T,T^*)R_i, S) \leq a$ and thus $T \in Y$.

THEOREM 2.6. Let $\mathfrak{F}$ denote the set of all factor operators on a separable Hilbert space $\mathcal{H}$. Then, relative to the $*$-strong operator topology, $\mathfrak{F}$ is a $F_{\sigma\delta\sigma}$ set in $\mathcal{L}(\mathcal{H})$ and $\mathfrak{F} \cap \mathcal{L}_1(\mathcal{H})$ is an $F_{\sigma\delta}$ set in $\mathcal{L}_1(\mathcal{H})$.

PROOF. Let d be a metric in $\mathcal{L}_1(\mathcal{H})$ which is compatible with the weak operator topology. Let $\mathcal{P}$ denote the collection of all polynomials in two noncommuting variables, with complex coefficients of the form $\alpha + \beta i$, where α and β are rational numbers. Then $\mathcal{P}$ is a countable set whose elements we enumerate $p_1, p_2, \ldots$. Further since $\mathcal{H}$ is separable, $\mathcal{L}_1(\mathcal{H})$ is weakly separable. Let $S_1, S_2, \ldots,$ denote a sequence of operators weakly dense in $\mathcal{L}_1(\mathcal{H})$. Let k, n, t, m denote positive integers. Let $Z_{k,n,t,m}$ denote the set of all operators T, $T \in \mathcal{L}_m(\mathcal{H})$ such that there exists operators $R_1, R_2, \ldots, R_n$ in $\mathcal{L}_n(\mathcal{H}) \cap \mathcal{A}(T)'$ such that

$$\sum_{i=1}^n p_i(T,T^*)R_i \in \mathcal{L}_1(\mathcal{H})$$

and

$$d\left(\sum_{i=1}^n p_i(T,T^*)R_i , S_k\right) \leq \frac{1}{t} .$$

Then by lemma 2.4 we have that each set $Z_{k,n,t,m}$ is $*$-strong closed. We shall show that $\mathfrak{F}$ is a $F_{\sigma\delta\sigma}$ set by proving

$$\mathfrak{F} = \bigcup_m \bigcap_{k,t} \bigcup_n Z_{k,n,t,m}$$

and hence that

$$\mathfrak{F} \cap \mathcal{L}_1(\mathcal{H}) = \bigcap_{k,t} \bigcup_n Z_{k,n,t,1}$$

is an $F_{\sigma\delta}$ set in $\mathcal{L}_1(\mathcal{H})$.

Suppose first that $T \in \bigcup_m \bigcap_{k,t} \bigcup_n Z_{k,n,t,m}$. Let $\mathcal{N}(T)$ denote the von Neumann algebra generated by $\mathcal{A}(T) \cup \mathcal{A}(T)'$ and let m be any positive integer, $m > \|T\|$. Then

$$T \in \bigcap_{k,t} \bigcup_n Z_{k,n,t,m}.$$

Let t and k denote any two positive integers. Then there exists a positive integer n such that

$$T \in Z_{k,n,t,m}.$$

Hence there exist operators $R_1, R_2, \ldots, R_n$ in $\mathcal{L}_n(\mathcal{H}) \cap \mathcal{A}(T)'$ such that

$$\sum_{i=1}^{n} p_i(T,T^*)R_i \in \mathcal{L}_1(\mathcal{H})$$

and

$$d\left(\sum_{i=1}^{n} p_i(T,T^*)R_i\, , S_k\right) \leq \frac{1}{t}.$$

Since $\sum_{i=1}^{n} p_i(T,T^*)R_i \in \mathcal{N}(T)$ we have

$$d(\mathcal{N}(T), S_k) \leq \frac{1}{t}.$$

Since this is true for any t and $\mathcal{N}(T)$ is weakly closed we have $S_k \in \mathcal{N}(T)$. Since this is also true for any k and $\mathcal{L}_1(\mathcal{H})$ is weakly closed we have

$\mathcal{L}_1(\mathcal{H}) \subset \mathcal{N}(T)$ and thus $\mathcal{N}(T) = \mathcal{L}(\mathcal{H})$. Thus the center of $\mathcal{a}(T)$ is the commutator of $\mathcal{N}(T)$, which is trivial, i.e., T is a factor operator.

Conversely, suppose T is a factor operator on $\mathcal{H}$. Choose an integer m such that $m > \|T\|$. Let k and t denote two positive integers. We shall complete the proof by showing that $T \in Z_{k,n,t,m}$. Let $C(T)$ denote the space of all finite sums of operators of the form $p(T,T^*)R$ where p is a polynomial in two noncommuting variables and $R \in \mathcal{a}(T)'$. Since T is a factor operator, the strong closure of $C(T)$ is $\mathcal{L}(\mathcal{H})$. By the Kaplansky density theorem ([89] or theorem 3, page 43 of [37]) there exists an operator P in $C(T)$ such that $\|P\| < 1$ and $d(P, S_k) < 1/t$. This P is of the form

$$P = \sum_{j=1}^{r} q_j(T,T^*)R_j$$

where the q_j are polynomials in two noncommuting variables and each $R_j \in \mathcal{a}(T)'$. For each j, $1 \leq j \leq r$, one may choose a polynomial p_{i_j} in $\mathcal{P}$ whose coefficients are so close to those of q_j that

$$\left\| \sum_{j=1}^{r} p_{i_j}(T,T^*)R_j \right\| \leq 1$$

and

$$d\left(\sum_{j=1}^{n} p_{i_j}(T,T^*)R_j, S_k \right) \leq \frac{1}{t}.$$

Now choose n larger than any of the numbers $i_1, i_2, \ldots, i_r$, $\|R_1\|, \ldots, \|R_r\|$. Then we have that $T \in Z_{k,n,t,m}$.

PROPOSITION 2.7. The set of normal operators is an $\mathcal{F}_\sigma$ set in $\mathcal{L}(\mathcal{H})$ relative to the $*$-strong operator topology.

PROOF. Note that, for each n, the set $\mathcal{N}_n$ of normal operators of norm less than or equal to n is $*$-strong closed by the use of proposition 2.2. Thus the set $\mathcal{N}$ of all normal operators is of the form $\mathcal{N} = \bigcup_{n=1}^{\infty} \mathcal{N}_n$ and hence an $\mathfrak{F}_\sigma$.

COROLLARY 2.8. The set of normal operators and the set of unitary operators are both Borel sets in $\mathcal{L}(\mathcal{H})$.

PROOF. The assertion for normal operators is just propositions 2.3 and 2.7. The assertion for unitary operators is given by lemma 1, page 166 of [37].

Edward Effros [46] has defined and studied a natural Borel structure on the collection $\mathcal{W}(\mathcal{L}(\mathcal{H}))$ of all σ-weakly closed subspaces of $\mathcal{L}(\mathcal{H})$ (and hence on the collection $\mathfrak{a}$ of von Neumann algebras defined on $\mathcal{H}$). The collection $\mathfrak{a}$ is a Borel subset of $\mathcal{W}(\mathcal{L}(\mathcal{H}))$ (corollary 1, page 1160 of [46]) and a standard Borel space. Further the maps $T \to \mathcal{A}(T)$ and $T \to \mathcal{A}(T)'$ of $\mathcal{L}(\mathcal{H})$ into $\mathfrak{a}$ are both Borel maps (cf. theorem 3 and also the remark in the first paragraph of page 1161, of [46]).

The reader may recall that Paul Halmos has proven that the set of irreducible operators is a dense G_δ in $\mathcal{L}(\mathcal{H})$ (cf. [79]). For clarity we remind the reader that this is relative to the norm topology and the following assertion refers to the Borel structure generated by the weak operator topology.

PROPOSITION 2.9. The set of irreducible operators is a Borel subset of $\mathcal{L}(\mathcal{H})$. Further the set of type I operators and the set of finite operators are Borel subsets of $\mathcal{L}(\mathcal{H})$. Further the sets of factor operators of each of the types I, I_∞, II_1, II_∞, III are Borel subsets of $\mathcal{L}(\mathcal{H})$.

PROOF. Since the collection $\mathfrak{A}$ of von Neumann algebras on $\mathcal{H}$ is a standard Borel space, singleton sets are Borel. Let $\mathcal{I}$ denote the trivial von Neumann algebra consisting of complex multiples of the identity operator. Then the singleton set $\{\mathcal{I}\}$ is a Borel set of $\mathfrak{A}$ and the inverse image, under the Borel map $T \to \mathcal{Q}(T)'$, is precisely the set of irreducible operators.

According to corollary 2.4, page 436 of [47] the set of type I von Neumann algebras is a Borel set. Further according to theorem 2.8 of [47] the set of all finite von Neumann algebras is a Borel subset of $\mathfrak{A}$. Since the map $T \to \mathcal{Q}(T)'$ is a Borel map of $\mathcal{L}(\mathcal{H})$ into $\mathfrak{A}$, proposition 1.38 implies that the set of type I operators and the set of finite operators are both Borel subsets of $\mathcal{L}(\mathcal{H})$.

Since both the set of factor operators (theorem 2.6) and the set of type I operators are Borel, so is their intersection. Thus also the finite type I factor operators form a Borel set and its relative complement in the type I factor operators, namely the factor operators of type I_∞ are also Borel. Now Ole Nielsen has shown (cf. theorem 3.6 of [100]) that the type III factor von Neumann algebras form a Borel subset of the set $\mathfrak{A}$ of all von Neumann algebras on the space $\mathcal{H}$, relative to the Effros Borel structure. Taking the inverse image under the Borel map $T \to \mathcal{Q}(T)$ we see that the set of type III factor operators form a Borel subset of $\mathcal{L}(\mathcal{H})$. Now a simple argument involving relative complements ensures that the factor operators of type II_1 and type II_∞ are Borel sets.

Incidentally, since the map $T \to \mathcal{Q}(T)$ is also a Borel map of $\mathcal{L}(\mathcal{H})$ into $\mathfrak{A}$, it follows that the class of operators considered by Carl Pearcy in [106], namely those operators generating finite type I von Neumann algebras, also form a Borel set in $\mathcal{L}(\mathcal{H})$. We also remark that the map $\mathcal{L}(\mathcal{H})$ into $\mathfrak{C}$ defined by $T \to \mathcal{A}(T) \cap \mathcal{A}(T)'$ is Borel by corollary 2, page 1160 of [46] and hence the inverse image of the singleton set $\{\mathcal{I}\}$, under this map, namely the set of factor operators in $\mathcal{L}(\mathcal{H})$, is also Borel (compare this with theorem 2.6).

We next turn our attention to the relation of unitary equivalence and quasi-equivalence to the Borel structure of $\mathcal{L}(\mathcal{H})$ and will eventually show that unitary equivalence classes and also quasi-equivalence classes are Borel sets in $\mathcal{L}(\mathcal{H})$.

If S and T are two operators on the same space $\mathcal{H}$, let $\mathcal{R}(S,T)$ denote the σ-weakly closed subspace of $\mathcal{L}(\mathcal{H})$ consisting of all intertwining transformations for S and T (cf. definition 1.4).

PROPOSITION 2.10. The map $\mathcal{L}(\mathcal{H}) \times \mathcal{L}(\mathcal{H}) \to \mathcal{V}(\mathcal{L}(\mathcal{H}))$ defined by $(S,T) \to \mathcal{R}(S,T)$ is Borel.

PROOF. (Cf. theorem 4 of [46] and theorem 2.8 of [94]) Here $\mathcal{V}(\mathcal{L}(\mathcal{H}))$ is the collection of σ-weakly closed linear subspaces of $\mathcal{L}(\mathcal{H})$, with the Borel structure given in [46]. Let $\mathcal{L}_*$ denote the separable Banach space of σ-weakly continuous complex valued linear functions on $\mathcal{L}(\mathcal{H})$. The $\mathcal{L}(\mathcal{H})$ can be identified as the dual of $\mathcal{L}_*$ in the canonical way, $\langle A,f\rangle = f(A)$ (cf. theorem 1, page 38 of [37]). Let $\mathcal{M}$ denote the Banach space $\mathcal{L}(\mathcal{H}) \times \mathcal{L}(\mathcal{H})$ with norm $\|(S,T)\| = \mathrm{Max}\left\{\|S\|, \|T\|\right\}$ and where the linear operations are defined componentwise. Let $\mathcal{M}_*$ denote the Banach space $\mathcal{L}_* \times \mathcal{L}_*$ with norm

$$\|(f_1,f_2)\| = \|f_1\| + \|f_2\|.$$

Then $\mathcal{M}$ may be identified with the dual of $\mathcal{M}_*$ by

$$\langle (S,T),\ (f_1,f_2)\rangle = f_1(S) + f_2(T).$$

For each pair of operators S and T in $\mathcal{L}(\mathcal{H})$, define the map $M^{(S,T)}$ of $\mathcal{L}(\mathcal{H})$ into $\mathcal{M}$ by

$$M^{(S,T)}(B) = (BS - TB,\ BS^* - T^*B).$$

Then the kernel of $M^{(S,T)}$ is $\mathcal{R}(S,T)$ and $M^{(S,T)}$ is σ-weakly continuous. Further $M^{(S,T)}$ is the dual of the map $M_*^{(S,T)} : \mathcal{M}_* \to \mathcal{L}_*$ defined by

$$\left[M_*^{(S,T)}(f,g)\right](B) = f(BS - TB) + g(BS^* - T^*B)$$

for all B in $\mathcal{L}(\mathcal{H})$ and $(f,g) \in \mathcal{M}_*$. Then the range of $M_*^{(S,T)}$ is $\left[\ker M^{(S,T)}\right]^\perp = [\mathcal{R}(S,T)]^\perp$. Thus if $\left\{(f_j,g_j) : j = 1,2,\ldots\right\}$ is a countable set which is norm dense in $\mathcal{M}_*$ then

$$\left\{M_*^{(S,T)}(f_j,g_j) : j = 1,2,\ldots\right\}$$

is σ-weak dense in $[\mathcal{R}(S,T)]^\perp$. Hence for any f in $\mathcal{L}_*$ we have

$$\|f + \mathcal{R}(S,T)^\perp\| = \operatorname{glb}\left\{\|f + M_*^{(S,T)}(f_j,g_j)\| : j = 1,2,\ldots\right\}.$$

However each of the functions

$$(S,T) \to \|f + M_*^{(S,T)}(f_j,g_j)\|$$

is Borel since each one is equal to

$$\operatorname{lub}\left\{|f(B_i) + f_j(B_iS - TB_i) + g_j(B_iS^* - T^*B_i)| : i = 1,2,\ldots\right\}$$

and further, for each fixed i, each of the functions

$$(S,T) \to |f(B_i) + f_j(B_iS - TB_i) + g_j(B_iS^* - T^*B_i)|$$

are Borel since each f_j and g_j are σ-weak continuous and hence Borel on

$\mathcal{L}(\mathcal{H})$ and for each fixed operator B_i, the maps $(S,T) \to B_iS - TB_i$ and $(S,T) \to B_iS^* - T^*B_i$ are σ-weak continuous maps of $\mathcal{L}(\mathcal{H}) \times \mathcal{L}(\mathcal{H})$ into $\mathcal{L}(\mathcal{H})$.

COROLLARY 2.11. The set $\{(S,T) : S \overset{|}{\circ} T\}$ is a Borel subset of $\mathcal{L}(\mathcal{H}) \times \mathcal{L}(\mathcal{H})$. If T is a fixed operator then $\{S : S \overset{|}{\circ} T\}$ is a Borel subset of $\mathcal{L}(\mathcal{H})$. If T is a factor operator, the quasi-equivalence class of T, $\{S : S \approx T\}$, is a Borel subset of $\mathcal{L}(\mathcal{H})$.

PROOF. Since $\mathcal{W}(\mathcal{L}(\mathcal{H}))$ is a standard Borel space, singleton sets, and in particular the singleton set consisting of the trivial subspace $\{0\}$, are Borel sets. By Schur's lemma (corollary 1.8) the inverse image of $\{0\}$ under the Borel map given in proposition 2.10 is precisely $\{(S,T) : S \overset{|}{\circ} T\}$. Intersecting this with the closed set $\mathcal{L}(\mathcal{H}) \times \{T\}$ and projecting into $\mathcal{L}(\mathcal{H})$ gives that $\{S : S \overset{|}{\circ} T\}$ is a Borel subset of $\mathcal{L}(\mathcal{H})$. Assuming T is a factor operator, the complement of this set, $\{S : S \text{ is not disjoint from } T\}$ is a Borel set. Intersecting this set with the Borel set of factor operators (theorem 2.6) and noting that if two factor operators are not disjoint they are quasi-equivalent we have that $\{S : S \approx T\}$ is a Borel set.

PROPOSITION 2.12. Let S and T denote two operators. Then there exists a unique projection E in $\mathcal{A}(S)'$ such that the restriction S_1 of S to the range of E is covered by T and the restriction S_2 of S to the range of $(I-E)$ is disjoint from T. Further the projection E is actually in the center of $\mathcal{A}(S)'$.

PROOF. If $T \overset{|}{\circ} S$ of course take $E = 0$. Otherwise choose E in $\mathcal{A}(S)'$ minimal with the property that S restricted to the range of $(I-E)$ is disjoint from T. It is easy to verify that $S_1 \overset{|}{\circ} S_2$ and hence by proposition 1.35, part 3, E is in the center of $\mathcal{A}(S)'$.

DEFINITION 2.13. If E is the projection given in proposition 2.12, we call S_1 the S-shadow of T and call E (denoted $C(S,T)$) the <u>S-shadow projection of</u> T. We call the decomposition $T = S_1 \oplus S_2$ of proposition 2.12 the <u>Lebesgue decomposition</u> of S relative to T because of the close analogy with the Lebesgue decomposition of measures.

COROLLARY 2.14. Let S_1 denote the S-shadow of T and let T_1 denote the T-shadow of S. Then $S_1 \approx T_1$.

PROPOSITION 2.15. If $\mathscr{H}_1$ and $\mathscr{H}_2$ are two separable Hilbert spaces, then the map $\mathscr{L}(\mathscr{H}_1) \times \mathscr{L}(\mathscr{H}_2) \to \mathscr{L}(\mathscr{H}_1)$ defined by $(S,T) \to C(S,T)$ is Borel.

PROOF. The proof is an almost immediate operator theoretic adaptation of the representation theoretic proof of theorem 5 of [46].

COROLLARY 2.16. If $\mathscr{H}_1$ and $\mathscr{H}_2$ are two separable Hilbert spaces, then the covering relation and the quasi-equivalence relations are Borel subsets of $\mathscr{L}(\mathscr{H}_1) \times \mathscr{L}(\mathscr{H}_2)$.

PROOF. Note that T covers S if and only if the S-shadow projection of T, $C(S,T)$, is the identity operator I on $\mathscr{H}_1$. Since $\mathscr{L}(\mathscr{H}_1)$ has a standard Borel structure the singleton set $\{I\}$ is a Borel subset of $\mathscr{L}(\mathscr{H}_1)$ and the inverse image of this Borel set, under the Borel map $(S,T) \to C(S,T)$

is precisely $\{(S,T) : T \} S\}$.

Further

$$\{(S,T) : S \approx T\} = \{(S,T) : T \} S\} \cap \{(S,T) : S \} T\}$$

is a Borel subset of $\mathcal{L}(\mathcal{H}_1) \times \mathcal{L}(\mathcal{H}_2)$.

COROLLARY 2.17. (Cf. proposition 2 of [50]) Let $\mathcal{H}$ denote a separable Hilbert space and let $\mathcal{B}$ denote a Borel subset of $\mathcal{L}(\mathcal{H})$. Then $\mathcal{B}^q = \{T : T \approx S$ for some S in $\mathcal{B}\}$ is an analytic set. Further if $\mathcal{B}$ intersects each quasi-equivalence class in at most one point, then $\mathcal{B}^q$ is a Borel set.

PROOF. We remark that the Borel structure of $\mathcal{L}(\mathcal{H}) \times \mathcal{L}(\mathcal{H})$ may either be described as that generated by the Borel rectangles or as the Borel structure generated by the product topology. The two methods of defining a Borel structure on $\mathcal{L}(\mathcal{H}) \times \mathcal{L}(\mathcal{H})$ are equivalent for standard Borel structures and the complete metrizable topologies generating those structures. Thus $\mathcal{B} \times \mathcal{L}(\mathcal{H})$ and $\{(R,S) : R \approx S\}$ are both Borel subsets of $\mathcal{L}(\mathcal{H}) \times \mathcal{L}(\mathcal{H})$ and hence their intersection, $\mathcal{S} = \{(R,S) : R \in \mathcal{B},\ S \in \mathcal{L}(\mathcal{H})$ and $R \approx S\}$ is a Borel subset of $\mathcal{L}(\mathcal{H}) \times \mathcal{L}(\mathcal{H})$. Then the projection π_2 onto the second coordinate is continuous and thus Borel and $\pi_2(\mathcal{S}) = \mathcal{B}^q$ is analytic. Furthermore if $\mathcal{B}$ intersects each quasi-equivalence class in at most one point, then the projection π_2 is a one-to-one Borel map of a standard Borel space $\mathcal{S}$ into a standard Borel space $\mathcal{L}(\mathcal{H})$. It therefore follows from theorem 3.2 of [96] that its image, $\mathcal{B}^q$, is a Borel set.

COROLLARY 2.18. If $\mathcal{H}_1$ and $\mathcal{H}_2$ are separable Hilbert spaces and T is an operator on $\mathcal{H}_1$, then

$t = \{S : S \in \mathcal{L}(\mathcal{H}_2)$ and $S \approx T\}$ is a Borel subset of $\mathcal{L}(\mathcal{H}_2)$. That is to say, the quasi-equivalence class of T, taken on any separable Hilbert space, is a Borel set.

PROOF. $(\{T\} \times \mathcal{L}(\mathcal{H}_2)) \cap \{(R,S) : R \approx S\} = \{T\} \times t$ is a Borel subset of $\mathcal{L}(\mathcal{H}_1) \times \mathcal{L}(\mathcal{H}_2)$ and hence a standard Borel space. The projection π_2 onto the second coordinate is one-to-one on $\{T\} \times t$ and hence its image, t, is a Borel subset of $\mathcal{L}(\mathcal{H}_2)$ (cf. theorem 3.2 of [96]).

The above corollary extends to arbitrary operators the observation we made earlier (corollary 2.11) for factor operators. We next employ the techniques of [47] to obtain the corresponding statement for unitary equivalence of operators. Here we do not bother with the complication of considering two Hilbert spaces since, unlike the case for quasi-equivalence, unitarily equivalent operators act on spaces of the same dimension.

PROPOSITION 2.19. Let T be an operator on the separable Hilbert space $\mathcal{H}$. Then the set $t = \{S : S \in \mathcal{L}(\mathcal{H})$ and $S \simeq T\}$ is a Borel subset of $\mathcal{L}(\mathcal{H})$.

PROOF. Note that the weak, strong and $*$-strong operator topologies all collapse into one topology when restricted to the group $\mathcal{U}$ of unitary operators on $\mathcal{H}$ and relative to this topology, $\mathcal{U}$ is a complete metrizable topological group (cf. also corollary 6.4 of [51]). Further $(\mathcal{U}, \mathcal{L}(\mathcal{H}))$ is a transformation group by defining the action of $\mathcal{U}$ by means of the map $\varphi : \mathcal{U} \times \mathcal{L}(\mathcal{H}) \to \mathcal{L}(\mathcal{H})$ where $\varphi(U,T) = UTU^*$ for $U \in \mathcal{U}$ and $T \in \mathcal{L}(\mathcal{H})$. The map φ is clearly Borel as all the algebraic operations are Borel. For each T in $\mathcal{L}(\mathcal{H})$, let $\mathcal{U}(T)$ denote the stabilizer subgroup of T in $\mathcal{U}$, i.e., $\mathcal{U}(T)$ is the group of those unitary operators U for which $T = UTU^*$. Thus

$\mathcal{U}(T) = \mathcal{U} \cap \mathcal{A}(T)'$. Since $\mathcal{A}(T)'$ is strongly closed, $\mathcal{U}(T)$ is a closed subgroup of $\mathcal{U}$. By lemma 3 of [38] there exists a Borel set B in $\mathcal{U}$, which intersects each left coset of $\mathcal{U}(T)$ in precisely one point. Then $B \times \{T\}$ is a Borel subset of the standard Borel space $\mathcal{U} \times \mathcal{L}(\mathcal{H})$ and hence is itself a standard Borel space. Since

$$\begin{aligned} t &= \{S : S \in \mathcal{L}(\mathcal{H}) \text{ and } S \cong T\} \\ &= \varphi(\mathcal{U} \times \{T\}) \\ &= \varphi(B \times \{T\}), \end{aligned}$$

t is the one-to-one Borel image of a standard Borel space and hence is itself a standard Borel space by theorem 3.2 of [96]. By corollary 1 to theorem 3.2 of [96] we have that t is a Borel subset of $\mathcal{L}(\mathcal{H})$.

2. The concrete dual of an operator

DEFINITION 2.20. For the sake of concreteness let $\mathcal{H}_n$ denote the n-dimensional Hilbert space of n-tuples of complex numbers and let $\mathcal{H}_0$ denote ℓ_2. Let $\mathcal{L} = \bigcup_{n=0}^{\infty} \mathcal{L}(\mathcal{H}_n)$. $\mathcal{L}$ is given a natural Borel structure by defining a subset B of $\mathcal{L}$ to be Borel if and only if $B \cap \mathcal{L}(\mathcal{H}_n)$ is a Borel subset of $\mathcal{L}(\mathcal{H}_n)$ for every n, $n = 0,1,2,\ldots$, relative to the Borel structure generated by the $*$-strong operator topology.

DEFINITION 2.21. Let T be an operator on a possibly nonseparable Hilbert space. The <u>concrete dual of</u> T, denoted T^c, is defined to be the subset of those elements of $\mathcal{L}$ which are weakly contained in T.

DEFINITION 2.22. (Cf. [56]) Let $\mathfrak{A}$ denote a C^*-algebra with identity. For each n, $n = 0,1,\ldots$, let $\mathfrak{A}_n^c$ denote the set of all

$*$-representations π of $\mathfrak{A}$ into $\mathcal{L}(\mathcal{H}_n)$ such that the identity element of $\mathfrak{A}$ is mapped into the identity operator on $\mathcal{H}_n$. $\mathfrak{A}_n^c$ is given the smallest topology such that, for each a in $\mathfrak{A}$, the map $\pi \to \pi(a)$ of $\mathfrak{A}_n^c$ into $\mathcal{L}(\mathcal{H}_n)$ is continuous relative to the weak operator topology of $\mathcal{L}(\mathcal{H}_n)$. (A straight-forward argument shows that the same (Fell) topology on $\mathfrak{A}_n^c$ is defined if one uses instead the $*$-strong operator topology in $\mathcal{L}(\mathcal{H}_n)$.) The <u>concrete dual</u> of $\mathfrak{A}$, denoted $\mathfrak{A}^c$, is defined to be $\mathfrak{A}^c = \bigcup_{n=0}^{\infty} \mathfrak{A}_n^c$. The concrete dual $\mathfrak{A}^c$ is a Borel space under the requirement that a set B, $B \subset \mathfrak{A}^c$ is Borel if and only if $B \cap \mathfrak{A}_n^c$ is a Borel subset of $\mathfrak{A}_n^c$ for every n, $n = 0,1,\dots$.

PROPOSITION 2.23. The concrete dual T^c of an operator T is a bounded (by $\|T\|$) Borel subset of $\mathcal{L}$. Indeed $T^c \cap \mathcal{L}(\mathcal{H}_n)$ is a $*$-strong closed subset of $\mathcal{L}(\mathcal{H}_n)$ for each n, $n = 0,1,2,\dots$. Further there is a natural Borel isomorphism φ of $[C^*(T)]^c$ onto T^c defined by $\varphi(\pi) = \pi(T)$. Indeed the restriction of φ to $[C^*(T)]_n^c$ is a $*$-strong homeomorphism onto $T^c \cap \mathcal{L}(\mathcal{H}_n)$, for each n, $n = 0,1,2,\dots$.

PROOF. Clearly every operator weakly contained in T has norm less than or equal $\|T\|$. For a fixed n we next show that $T^c \cap \mathcal{L}(\mathcal{H}_n)$ is $*$-strong closed in $\mathcal{L}(\mathcal{H}_n)$. Let S_λ, $\lambda \in \Lambda$, denote a net in $T^c \cap \mathcal{L}(\mathcal{H}_n)$ such that S_λ converges to S in the $*$-strong operator topology, where $S \in \mathcal{L}(\mathcal{H}_n)$. By the definition of weak containment there exists a net π_λ of elements of $[C^*(T)]_n^c$ such that $S_\lambda = \pi_\lambda(T)$ for each λ in Λ. Since π_λ is a $*$-representation we have

$$\pi_\lambda(p(T,T^*)) = p(S_\lambda, S_\lambda^*)$$

for every complex polynomial p in two noncommuting variables (cf. definition 1.31). Now all the S_λ and S_λ^* are bounded by $\|T\|$. Since on bounded sets both multiplication and the adjoint operation are $*$-strong continuous (proposition 2.2) we have that $p(S_\lambda, S_\lambda^*)$ converges to $p(S,S^*)$ in the $*$-strong operator topology.

We next proceed to define a $*$-representation π of $C^*(T)$ into $\mathcal{L}(\mathcal{H}_n)$ as follows. If $p(T,T^*)$ is any element of the polynomial algebra of T (cf. definition 1.31) we define $\pi(p(T,T^*)) = p(S,S^*)$. To see that π is well defined on the polynomial algebra of T, suppose q is a second complex polynomial in two noncommuting variables such that $p(T,T^*) = q(T,T^*)$. Then, (using limits relative to the $*$-strong topology of $\mathcal{L}(\mathcal{H}_n)$) we have

$$\begin{aligned} p(S, S^*) &= \lim_\lambda p(S_\lambda, S_\lambda^*) \\ &= \lim_\lambda \pi_\lambda(p(T, T^*)) \\ &= \lim_\lambda \pi_\lambda(q(T, T^*)) \\ &= \lim_\lambda q(S_\lambda, S_\lambda^*) \\ &= q(S, S^*). \end{aligned}$$

Thus π is well defined and, since $\pi(T) = S$, preserves the $*$-algebra operations of the polynomial algebra of T, i.e., $\pi(p(T,T^*)) = p(S,S^*) = p(\pi(T), \pi(T)^*)$. To see that π can now be extended uniquely to a $*$-representation of $C^*(T)$ it is sufficient to show that π is norm non-increasing on the polynomial algebra $\mathcal{P}(T)$ of T which is norm dense in $C^*(T)$. While the norm is not continuous relative to the operator topologies, there is a kind of semi-continuity which just manages to do the trick (cf. §29 of [114]). Indeed since $\pi_\lambda(p(T,T^*))$ converges strongly (and hence weakly) to $p(S,S^*)$ it follows that

$$\|\pi(p(T,T^*))\| = \|p(S,S^*)\| \leq \liminf_\lambda \|\pi_\lambda(p(T,T^*))\| \leq \|p(T,T^*)\| .$$

Thus π can be extended to a $*$-representation of $C^*(T)$ and since $\pi(T) = S$ we have $S \in T^c \cap \mathcal{L}(\mathcal{H}_n)$.

We next verify that the map φ of $[C^*(T)]_n^c$ into $T^c \cap \mathcal{L}(\mathcal{H}_n)$ is a $*$-strong homeomorphism. As we have mentioned in definition 2.22, if a net π_λ of $*$-representations in $[C^*(T)]_n^c$ converges to another $*$-representation in the Fell topology for $[C^*(T)]_n^c$ then $\pi_\lambda(T) = \varphi(\pi_\lambda)$ converges to $\pi(T) = \varphi(\pi)$ in the $*$-strong operator topology. Conversely if $\pi_\lambda(T) \to \pi(T)$ in the $*$-strong operator topology, proposition 2.2 implies $\pi_\lambda(p(T,T^*))$ converges to $\pi(p(T,T^*))$ in the $*$-strong operator topology for each element $p(T,T^*)$ of the polynomial algebra of T, which algebra is norm dense in $C^*(T)$. Thus $\pi_\lambda(S)$ converges to $\pi(S)$ in the $*$-strong operator topology for each S in $C^*(T)$. Hence π_λ converges to π in the Fell topology of $[C^*(T)]_n^c$.

DEFINITION 2.24. Let T be an operator on a not necessarily separable Hilbert space. Let T^i (respectively T^f, T^I) denote the set of all irreducible (respectively factor, type I) operators in $\mathcal{L}$ which are weakly contained in T.

COROLLARY 2.25. The subsets T^i, T^f, T^I and $T^f \cap T^I$ are all Borel subsets of the standard Borel space T^c and hence are themselves standard Borel spaces.

PROOF. This is a direct application of theorem 2.6, proposition 2.9 of this memoir and corollary 1, page 139 of [96].

3. The spectrum and quasi-spectrum of an operator

DEFINITIONS 2.26. Consider the set $\mathcal{L}$ of operators given by definition 2.20 which concretely contains representatives of all possible bounded linear operators on separable Hilbert spaces, both finite dimensional and infinite dimensional. Let $\mathfrak{F}$ denote the subset of $\mathcal{L}$ consisting of all factor operators in $\mathcal{L}$ and let $\mathcal{I}$ denote the subset of $\mathfrak{F}$ consisting of all irreducible operators in $\mathcal{L}$. By propositions 2.3, 2.9 and theorem 2.6 of this memoir, as well as theorem 3.1 and corollary 1 to theorem 3.2 of [96], $\mathcal{I}$ and $\mathfrak{F}$ are Borel subsets (and hence standard Borel spaces) of the standard Borel space $\mathcal{L}$. Let $\tilde{\mathcal{I}} \subset \tilde{\mathfrak{F}} \subset \tilde{\mathcal{L}}$ denote the quotient spaces of $\mathcal{I}$, $\mathfrak{F}$ and $\mathcal{L}$, respectively, with respect to quasi-equivalence. We shall call $\tilde{\mathfrak{F}}$ the <u>quasi-factor space</u>. Note that $\tilde{\mathcal{I}}$ might also be characterized as the space of unitary equivalence classes of irreducible operators acting on separable Hilbert spaces, since irreducible operators are quasi-equivalent if and only if they are unitarily equivalent. (This is clear, for example, from corollary 1.14.) Alternatively corollary 1.44 enables one to identify $\tilde{\mathcal{I}}$ as the type I part of $\tilde{\mathfrak{F}}$, i.e., $\tilde{\mathcal{I}}$ is the space of quasi-equivalence classes of type I factor operators.

We give $\tilde{\mathcal{L}}$ the quotient Borel structure, i.e., the largest Borel structure such that the quotient map $r : \mathcal{L} \to \tilde{\mathcal{L}}$ is Borel. Thus a set B contained in $\tilde{\mathcal{L}}$ is Borel if and only if $r^{-1}(B)$ is a Borel subset of $\mathcal{L}$. Thus in particular $\tilde{\mathfrak{F}}$ and $\tilde{\mathcal{I}}$ are Borel subsets of $\tilde{\mathcal{L}}$. Of course both $\tilde{\mathfrak{F}}$ and $\tilde{\mathcal{I}}$ may be given a quotient Borel structure and it is easy to verify that this gives rise to the same Borel structure they acquire as subspaces of $\tilde{\mathcal{L}}$. This quotient Borel structure is also called the Mackey Borel structure.

PROPOSITION 2.27. The space $\tilde{\mathcal{L}}$ (and hence the subspaces $\tilde{\mathfrak{F}}$ and $\tilde{\mathcal{I}}$) are separated Borel spaces. Indeed singleton sets are always Borel.

PROOF. This follows immediately from corollary 2.18.

The goal of this memoir is to show how the unitary equivalence problem for all operators on a separable space can be reduced (at least for the broad-minded reader) to the problem of determining the quasi-factor space $\tilde{\mathfrak{F}}$ and the unitary equivalence problem for type I operators can be reduced to determining the Borel space $\tilde{\mathcal{I}}$ of irreducible operators divided by unitary equivalence. From this grandiose point of view we mathematical creatures must accept the humbling news (corollary 2.33 below) that neither $\tilde{\mathfrak{F}}$ nor $\tilde{\mathcal{I}}$ are standard Borel spaces and in fact are not even countably separated. (In the terminology of George Mackey we say $\tilde{\mathfrak{F}}$ and $\tilde{\mathcal{I}}$ are not smooth.) Thus it is impossible to describe either $\tilde{\mathfrak{F}}$ or $\tilde{\mathcal{I}}$ by means of countably many Borel invariants. By now such bad news, while disheartening, is not surprising. For example recently James Woods [133] has shown that the classification of factor von Neumann algebras is also not smooth. While it may have been ordained that we shall never classify all operators up to unitary equivalence before the sun burns out, there is no reason why we presumptuous primates should not continue to explore the entangled jungles $\tilde{\mathfrak{F}}$ and $\tilde{\mathcal{I}}$, and even to colonize many of their smooth territories. And so we continue.

DEFINITION 2.28. Let T be an operator. By the quasi-spectrum of T, denoted $\tilde{T}$, we mean the space of quasi-equivalence classes of factor operators weakly contained in T. By the <u>spectrum</u> of T, denoted $\hat{T}$, we mean the space of unitary equivalence classes of irreducible operators weakly contained in T.

Under the identifications given in definitions 2.26, we have $\hat{T} \subset \tilde{\mathcal{J}} \subset \tilde{\mathfrak{F}}$ and $\tilde{T} \subset \tilde{\mathfrak{F}}$. Indeed $\hat{T} = \tilde{T} \cap \tilde{\mathcal{J}}$. Thus we may consider the spectrum of T to be the type I part of the quasi-spectrum of T. Thus it is clear from proposition 1.53 ($1 \Leftrightarrow 2$) that T is smooth if and only if $\hat{T} = \tilde{T}$.

PROPOSITION 2.29. If T is an operator, then $\hat{T}$ and $\tilde{T}$ are both nonempty Borel subsets of $\tilde{\mathfrak{F}}$ and hence have Borel structures equivalently defined either as Borel subspaces of $\tilde{\mathfrak{F}}$, or as quotient spaces of $\mathcal{J} \cap T^c$ and $\mathfrak{F} \cap T^c$ respectively.

PROOF. This follows from the fact that $\mathcal{J}$, $\mathfrak{F}$ and T^c are all Borel subsets of $\mathcal{L}$ (cf. proposition 2.23, definitions 2.26 and corollary 2.25). Further $\hat{T}$ is nonempty since $C^*(T)$ admits an irreducible $*$-representation φ and hence the unitary equivalence class of $\varphi(T)$ is in $\hat{T}$.

REMARK 2.30. A few remarks on terminology are in order to justify the arrogance of associating a new object, $\hat{T}$, with an operator, and calling it the "spectrum" of the operator. Classically the "spectrum" of an operator T is that compact subset of the complex plane consisting of those complex numberts λ for which $(T - \lambda I)$ is not invertible. Throughout this memoir this set will be denoted $\Lambda(T)$ and will be called the <u>numerical spectrum</u> of T. If T is a normal operator it is easy to verify that all the irreducible operators weakly contained in T act on a one-dimensional space. (Since T is normal, $C^*(T)$ is abelian and hence if S is an irreducible operator weakly contained in T, $\mathcal{A}(S)$ is abelian and thus $\mathcal{A}(S) \subset \mathcal{A}(S)'$ which consists only of complex multiples of the identity operator since S is irreducible. Thus S is of the form λI where $\lambda \in \mathbb{C}$ and I is the identity operator. But since S is irreducible I must act on the one-dimensional space $\mathcal{H}_1 = \mathbb{C}$.) Thus to each element in $\hat{T}$ we associate a complex number in $\mathbb{C}$, and under this identification $\hat{T} = \Lambda(T)$. Further the Borel structure we have defined on $\hat{T}$ corresponds to the ordinary Borel structure of $\Lambda(T)$ as a subset of the complex plane. This is not a new observation but is implicit in any proof of the spectral theorem for normal operators based on the Gelfand-Naimark theorem (cf. for example theorems 4.29 and 4.30 of [42]). For nonnormal operators the one-dimensional part of $\hat{T}$ is always a proper Borel subset of $\hat{T}$, called the <u>normal spectrum of</u> T (which can in fact be empty), and may be identified with a closed subset of $\Lambda(T)$.

Another justification for this terminology is that it is consistent with well-established terminology for noncommutative C^*-algebras (cf. [36]), which will be made explicit in corollary 2.31 below. However the real justification for this terminology, the validity of which the reader may judge after reading this monograph, is that the spectrum $\hat{T}$ plays precisely the same role in a general "spectral theorem" for smooth operators, that the numerical spectrum plays in the spectral theorem for normal operators. That is a smooth operator T is decomposed relative to a finite Borel measure μ on its spectrum $\hat{T}$, and the operator T is determined, up to unitary equivalence, by its spectrum $\hat{T}$, the measure class of μ on $\hat{T}$, and a spectral multiplicity function defined on the measure classes absolutely continuous with respect to μ. If T is not smooth, so that $\hat{T} \neq \tilde{T}$, the same remarks hold except that it is the quasi-spectrum $\tilde{T}$ which plays the feature role and the notion of spectral multiplicity function must be generalized beyond that presented in [77].

We remark that the notion of quasi-equivalence for $*$-representations of a C^*-algebra $\mathfrak{A}$ is similar to the notion we have defined for operators (cf. proposition 5.3.1 and definition 5.3.2 of [36]). Further in the natural map φ which we have defined in proposition 2.23 of this memoir, mapping $[C^*(T)]^c$ onto T^c by $\varphi(\pi) = \pi(T)$, we have that two representations π_1 and π_2 in $[C^*(T)]^c$ are quasi-equivalent (respectively unitarily equivalent) if and only if $\varphi(\pi_1) = \pi_1(T)$ and $\varphi(\pi_2) = \pi_2(T)$ are quasi-equivalent (respectively unitarily equivalent). Further a representation π in $[C^*(T)]^c$ is a factor representation (respectively irreducible representation) if and only if $\pi(T)$ is a factor operator (respectively an irreducible operator). The <u>spectrum</u> of a C^*-algebra $\mathfrak{A}$ is defined to be the Borel space $\hat{\mathfrak{A}}$ consisting of the set of unitary equivalence classes of irreducible $*$-representations in $\mathfrak{A}^c$, endowed with the quotient Borel structure (cf. definition 3.1.5, page 60 and definition 3.8.2, page 78 of [36]). The <u>quasi-spectrum</u> of a C^*-algebra $\mathfrak{A}$ is defined to be the Borel space $\tilde{\mathfrak{A}}$ consisting of all quasi-equivalence classes of factor $*$-representations in $\mathfrak{A}^c$, again endowed with the quotient Borel structure (cf. definition 7.2.2 page 136 of [36]). This concept was first introduced under the term <u>quasi-dual</u> (cf. page 254 and section 2 of [50]). These remarks are sufficient to establish the following corollary of proposition 2.23.

> COROLLARY 2.31. If T is an operator, acting on a possibly nonseparable space, then there exists a canonical Borel isomorphism of the spectrum $[C^*(T)]^\wedge$ and the quasi-spectrum $[C^*(T)]^\sim$ onto the spectrum $\hat{T}$ and quasi-spectrum $\tilde{T}$, of T, respectively, which associates with each equivalence

class of a $*$-representation π of $C^*(T)$, the corresponding equivalence class of the operator $\pi(T)$.

We have earlier lamented the fact that the space $\tilde{\mathfrak{F}}$ of quasi-equivalence classes of all factor operators on a separable Hilbert space, or even the subspace $\tilde{\mathcal{J}}$ of unitary equivalence classes of irreducible operators, are not smooth in the sense that they are not countably separated Borel spaces. Fortunately there is a large class of operators whose spectra and quasi-spectra have very well behaved Borel structures (which we call smooth) and this is in fact the real justification for the terminology introduced in definition 1.50 (cf. also propositions 1.51 and 1.53).

THEOREM 2.32. (Glimm [69]) Let T be an operator. Then the following conditions are equivalent.

1. T is a smooth operator
2. The quasi-spectrum $\tilde{T}$ is a standard Borel space.
3. The quasi-spectrum $\tilde{T}$ has a countable family of Borel sets which separate points.
4. The spectrum $\hat{T}$ is a standard Borel space.
5. The spectrum $\hat{T}$ has a countable family of Borel sets which separate points.
6. $\hat{T} = \tilde{T}$.

PROOF. This is simply theorem 2 of [69], transplanted into operator theory by corollary 2.31 and mildly mixed with propositions 1.51 and 1.53 above (cf. also [41]).

Note the rather sharp demarcation line that theorem 2.32 establishes between smooth and non-smooth operators. Either the spectrum $\hat{T}$ is very nice (either countable or Borel isomorphic to the unit interval with the Borel structure generated by the usual topology) or its Borel structure is not even countably separated.

COROLLARY 2.33. The space $\tilde{\mathfrak{F}}$ of all quasi-equivalence classes of factor operators acting on a separable Hilbert space, as well as the space $\tilde{\mathcal{J}}$ of all unitary equivalence classes of irreducible operators, do not admit a countable family of Borel sets which separate points.

PROOF. (Cf. remarks following proposition 2.27) Clearly if $\tilde{\mathfrak{F}}$ is countably separated then every subspace of $\tilde{\mathfrak{F}}$ determined by a Borel set is also countably separated. Yet for any operator T, we have $\hat{T} \subset \tilde{\mathcal{J}} \subset \tilde{\mathfrak{F}}$ and both $\hat{T}$ and $\tilde{\mathcal{J}}$ are Borel subsets of $\tilde{\mathfrak{F}}$ (cf. definitions 2.26 and proposition 2.29). Thus it is sufficient to find one operator T for which $\hat{T}$ is not countably separated as a Borel space. But by theorem 2.32 we need only exhibit a non-smooth operator and proposition 1.53 does just that.

REMARK 2.34. The Topping operator T of proposition 1.52 has the remarkable property that it is irreducible and yet both its spectrum $\hat{T}$ and quasi-spectrum $\tilde{T}$ are not countably separated as Borel spaces. Furthermore $\tilde{T}$ actually contains uncountably many type III factor operators. This is because the operator T generates a uniformly hyperfinite C^*-algebra, $C^*(T)$, and Robert Powers [111] has shown that such a C^*-algebra admits a continuum of mutually disjoint type III factor $*$-representations. The images of T under these representations, of course, give a continuum of type III elements of $\tilde{T}$. For a concrete realization of the Topping operator see the remark following corollary 8 of [23].

4. Some spectral theory

As Bill Arveson has pointed out (theorem 1.1.2 of [3]), two normal operators are weakly equivalent (algebraically equivalent in Arveson's terminology) if and only if they have the same spectrum. The following result generalizes this to arbitrary operators (see also corollary 2.38).

PROPOSITION 2.35. Let S and T denote two operators. Then the following three statements are equivalent.

1. S and T are weakly equivalent.
2. $\hat{S} = \hat{T}$.
3. $\tilde{S} = \tilde{T}$.

PROOF. It is clear from the definitions that 1 implies 2 and 3. Furthermore 3 implies 2 since $\hat{S} = \tilde{S} \cap \mathcal{J}$ and $\hat{T} = \tilde{T} \cap \tilde{\mathcal{J}}$. Thus it remains to prove that 2 implies 1. To each $t \in \hat{T}$, let T_t denote an irreducible operator in the equivalence class t. For each such operator let π_t denote the irreducible $*$-representation of $C^*(T)$ such that $\pi_t(T) = T_t$. (We know π_t exists because T_t is weakly contained in T.) Then $\pi = \sum_{t \in \hat{T}} \oplus \pi_t$ is a faithful $*$-representation of $C^*(T)$ onto $C^*\left(\sum_{t \in \hat{T}} \oplus T_t\right)$ by section 2.7.3, page 41 of [36]. Similarly there is a faithful $*$-representation of $C^*(S)$ onto $C^*\left(\sum_{t \in \hat{T}} \oplus T_t\right)$, since $\hat{S} = \hat{T}$. Combining these two $*$-isomorphisms we obtain a $*$-isomorphism φ of $C^*(S)$ onto $C^*(T)$ such that $\varphi(S) = T$. By proposition 1.47 we have $S \sim T$.

It is well known that if T is a normal operator, then its spectral radius, $\sup\{|\lambda| : \lambda \in \Lambda(T)\}$, is equal to $\|T\|$ (cf. for example, theorem 2, page 55 of [77]). The following proposition indicates this phenomenon persists for general operators, providing one uses our generalized notion of the spectrum of an operator.

PROPOSITION 2.36. For any operator T, there exists an irreducible operator T_0 weakly contained in T such that $\|T_0\| = \|T\|$. Thus

$$\|T\| = \operatorname{Sup}\{\|S\| : S \in \hat{T}\} = \operatorname{Max}\{\|S\| : S \in \hat{T}\}.$$

PROOF. According to lemma 3.3.6, page 64 of [36] the function on $[C^*(T)]^\wedge$ defined by $\pi \to \|\pi(T)\|$ attains its upper bound $\|T\|$. But under the identification of $[C^*(T)]^\wedge$ with $\hat{T}$ given by corollary 2.31, the function in question simply corresponds to the norm function, $S \to \|S\|$, on $\hat{T}$.

If p is a complex polynomial in two noncommuting variables we may again apply lemma 3.3.6, page 64 of [36] to the function on $[C^*(T)]^\wedge$ defined by $\pi \to \|\pi(p(T,T^*)\|$, which therefore attains its upper bound, $\|p(T,T^*)\|$. Thus essentially the same argument as we used for the previous proposition gives us the generalization of theorem 3, page 55 of [77], which was proved for Hermitian operators.

PROPOSITION 2.37. If T is an operator and p is a complex polynomial in two noncommuting unknowns, then

$$\begin{aligned}\|p(T,T^*)\| &= \operatorname{Sup}\{\|p(S,S^*)\| : S \in \hat{T}\} \\ &= \operatorname{Max}\{\|p(S,S^*)\| : S \in \hat{T}\}.\end{aligned}$$

The proof of theorem 3, page 55 of [77], involved an application of the polynomial mapping theorem for the spectrum of a Hermitian operator. Unfortunately we do not have a polynomial mapping theorem for the general spectrum $\hat{T}$, but the existence of proposition 2.37 indicates that some aspects of the phenomenon persist in the general situation.

COROLLARY 2.38. If S and T are operators, then S is weakly contained in T if and only if $\tilde{S} \subset \tilde{T}$.

PROOF. Since weak containment is transitive, it is obvious that if S is weakly contained in T, then $\tilde{S} \subset \tilde{T}$. Conversely suppose S is not weakly contained in T. Then there is a polynomial p in two noncommuting variables such that $\|p(S,S^*)\| > \|p(T,T^*)\|$. (Otherwise one could construct a $*$-homomorphism of the $*$-algebra generated by T onto the $*$-algebra generated by S, mapping T into S, which could be extended, by continuity to their norm closures, the C^*-algebras generated by T and S respectively.) By proposition 2.37 there is an irreducible operator S_0 whose unitary equivalence class is an element of $\tilde{S}$ and for which $\|p(S_0,S_0^*)\| = \|p(S,S^*)\| > \|p(T,T^*)\|$. Thus S_0 cannot be weakly contained in T and hence $\tilde{S}$ is not contained in $\tilde{T}$.

While we have lost the polynomial mapping theorem, some basic properties of transforms of spectra persist.

LEMMA 2.39. If T is an invertible operator then $C^*(T) = C^*(T^{-1})$. Further every operator S weakly contained in T is invertible and S^{-1} is weakly contained in T^{-1}.

PROOF. If T is an invertible operator, then T^{-1} is contained in $C^*(T)$ (cf. the proof of proposition 1.3.10, page 8 of [36]). Thus $C^*(T^{-1}) \subset C^*(T)$. By symmetry between an operator and its inverse we have $C^*(T^{-1}) = C^*(T)$.

If S is weakly contained in T, then there exists a $*$-homomorphism φ of $C^*(T)$ onto $C^*(S)$ such that $\varphi(T) = S$ and further the identity of $C^*(T)$ is mapped into the identity of $C^*(S)$. Thus $\varphi(T^{-1})$ is the inverse

S^{-1} of S. Thus φ maps $C^*(T^{-1})$ onto $C^*(S^{-1})$ and $\varphi(T^{-1}) = S^{-1}$, i.e., S^{-1} is weakly contained in T^{-1}.

COROLLARY 2.40. Let T be an invertible operator. Then any operator weakly equivalent to T (and hence in particular any operator quasi-equivalent to T) is also invertible. Furthermore T is irreducible (respectively a factor operator) if and only if T^{-1} is irreducible (respectively a factor operator).

The following proposition, an immediate consequence of lemma 2.39 and corollary 2.40, is a generalization of theorem 2, page 54 of [77].

PROPOSITION 2.41. If T is an invertible operator, then

$$(T^{-1})^{\wedge} = (T^{\wedge})^{-1}$$

and

$$(T^{-1})^{\sim} = (T^{\sim})^{-1}.$$

Similar facts hold, and are more obvious, for the adjoint of an operator. Thus for any operator T, $C^*(T) = C^*(T^*)$ and an operator S is weakly contained in T if and only if S^* is weakly contained in T^*. Further an operator S is irreducible (respectively a factor operator) if and only if S^* is irreducible (respectively a factor operator). Thus we have the following generalization of theorem 3, page 54 of [77].

PROPOSITION 2.42. If T is an operator, then

$$(T^*)^{\wedge} = (T^{\wedge})^*$$

and

$$(T^*)^{\sim} = (T^{\sim})^*.$$

PROPOSITION 2.43. If T is an operator, then its quasi-spectrum $\tilde{T}$ is the union of three mutually disjoint Borel sets,

$$\tilde{T} = \hat{T} \cup \tilde{T}_{II} \cup \tilde{T}_{III}$$

where $\tilde{T}_{II}$ (respectively $\tilde{T}_{III}$) consists of quasi-equivalence classes of type II (respectively type III) factor operators.

PROOF. This follows immediately from proposition 2.9 and corollary 2.25.

The above partitioning of the quasi-spectrum of an operator enables one to formulate still another characterization of smooth operators.

PROPOSITION 2.44. Let T be an operator and $\tilde{T} = \hat{T} \cup \tilde{T}_{II} \cup \tilde{T}_{III}$ its quasi-spectrum. Then the following three statements are equivalent.

1. T is a smooth operator.
2. $\tilde{T}_{II}$ is empty.
3. $\tilde{T}_{III}$ is empty.

PROOF. The fact that 1 implies 2 and 3 and the fact that 2 and 3, together, imply 1, has already been observed (for example theorem 2.32, $1 \Leftrightarrow 6$). However the equivalence of 2 and 3 is just the equivalence of statements a3 and a4 of theorem 1 of [69], properly transplanted into operator theory via corollary 2.31.

In an earlier remark (following proposition 2.27) we bemoaned the fact (corollary 2.33) that the quasi-factor space $\tilde{\mathfrak{F}}$ of all quasi-equivalence classes of factor operators on a separable Hilbert space is not smooth, in the sense that its Borel structure is not countably separated. However we consoled ourselves with the hope that there are nevertheless significant portions of $\tilde{\mathfrak{F}}$ which are smooth. We next note that it follows from some fundamental work of Alain Guichardet [74] that the subset $\tilde{\mathcal{G}}$ of $\tilde{\mathfrak{F}}$ corresponding to factor operators which generate finite von Neumann algebras (of either type I or II) forms a Borel subset of $\tilde{\mathfrak{F}}$ and further $\tilde{\mathcal{G}}$ is smooth, i.e., is a standard Borel space. (In particular $\tilde{\mathcal{G}}$ includes all the finite dimensional factor operators.) We have already defined an operator T to be finite if its <u>commuting</u> algebra is a finite von Neumann algebra (cf. proposition 1.38). We are thus led to introduce the following somewhat artificial terminology.

DEFINITION 2.45. A operator T will be called <u>cofinite</u> if the von Neumann algebra $\mathcal{Q}(T)$ which it generates is a finite von Neumann algebra.

PROPOSITION 2.46. The set $\tilde{\mathcal{G}}$ of quasi-equivalence classes of cofinite factor operators is a Borel subset of $\tilde{\mathfrak{F}}$ and is a standard Borel space.

PROOF. For each positive integer n, let $T(n)$ denote the operator which is the direct sum of all factor operators T acting on ℓ_2 such that $\|T\| \leq n$. Then for each n, the quasi-spectrum $\tilde{T}(n)$ of $T(n)$ is a Borel subspace of $\tilde{\mathfrak{F}}$ and $\tilde{\mathfrak{F}} = \bigcup_n \tilde{T}(n)$. Now if we translate the result of Alain Guichardet for representations of C^*-algebras (theorem 1, page 21 of [74] or proposition 7.4.3, p. 140 of [36]) into operator theoretic terms by means of corollary 2.31 we have that that portion of $\tilde{T}(n)$ corresponding to cofinite factor operators (i.e., $\tilde{T}(n) \cap \tilde{\mathcal{G}}$) is a Borel subset of $\tilde{T}(n)$ and is a standard Borel space. Thus $\tilde{\mathcal{G}} = \bigcup_n (\tilde{T}(n) \cap \tilde{\mathcal{G}})$ is a Borel subset of $\tilde{\mathfrak{F}}$

and, by theorem 3.1 of [96] $\tilde{\mathcal{G}}$ is a standard Borel space.

COROLLARY 2.47. The subset $\tilde{\mathcal{J}} \cap \tilde{\mathcal{G}}$ of $\tilde{\mathcal{J}}$, consisting of all unitary equivalence classes of cofinite irreducible operators, is a standard Borel space.

We remark that this smooth portion of $\tilde{\mathcal{J}}$ has already been explored and charted by Carl Pearcy [106].

COROLLARY 2.48. For any operator T, the cofinite part $\tilde{T}_g$ of its quasi-spectrum $\tilde{T}$, and the cofinite part $\hat{T}_g$ of its spectrum $\hat{T}$ are standard Borel spaces. (Here $\tilde{T}_g = \tilde{T} \cap \tilde{\mathcal{G}}$ and $\hat{T}_g = \hat{T} \cap \tilde{\mathcal{G}}$).

Of course, for each positive integer n, the set of irreducible operators acting on an n-dimensional space forms a Borel subset of $\tilde{\mathcal{J}}$ and thus, for any operator T, the n-dimensional part of $\hat{T}$ is a Borel subset of the cofinite part $\hat{T}_g$ of $\hat{T}$ and hence is itself an interesting smooth part of $\hat{T}$. Of particular interest is the standard Borel subspace of the spectrum $\hat{T}$ corresponding to one-dimensional irreducible operators weakly contained in T. This may clearly be identified in a natural manner with a bounded (by $\|T\|$) subset of the complex plane. In fact this one-dimensional part of $\hat{T}$ has already received extensive study in the last few years (cf. [5], [12], [17], [18], [49], [61], [62], [63], [64], [65], [79], [83], [84], [90], [92], [120]) under various names, such as, the set of normal approximate propervalues of T [90], the set of approximate reducing propervalues of T [120], the normal approximate spectrum [49]. We shall simply call this one-dimensional part of $\hat{T}$ (considered as a bounded subset of the complex plane) the <u>normal spectrum</u>

of T, and will denote it (following [49]) by $\Pi_n(T)$. This notation and terminology is justified as $\Pi_n(T)$ is a closed (and hence compact) subset of the approximate point spectrum $\Pi(T)$, and if T is normal, $\Pi_n(T)$ is equal to the ordinary spectrum $\Lambda(T)$ of T. The following proposition gathers together some of the various known characterizations of this one-dimensional part of $\hat{T}$ (see also theorem 1 of [92]).

PROPOSITION 2.49. Let T be an operator and let λ be a complex number. Then the following statements are equivalent.

1. $\lambda \in \Pi_n(T)$ the normal spectrum of T.
2. There exists a character φ defined on $C^*(T)$ such that $\varphi(T) = \lambda$.
3. For each $\epsilon > 0$, there exists a unit vector x in $\mathscr{K}(T)$ such that $\|(T - \lambda I)x\| < \epsilon$ and $\|(T - \lambda I)^* x\| < \epsilon$.
4. The closure of the numerical range of the operator
$$(T - \lambda I)^*(T - \lambda I) + (T - \lambda I)(T - \lambda I)^*,$$
contains zero.
5. The C^*-algebra generated by $(T - \lambda I)$ does not contain the identity operator.
6. The closed two-sided ideal generated by $(T - \lambda I)$ and $(T - \lambda I)^*$ in $C^*(T)$, the C^*-algebra generated by T and the identity operator, is a proper ideal of $C^*(T)$.

PROOF. Statements 1 and 2 are equivalent as the term character is an alias for one-dimensional irreducible $*$-representation. The implication 3 implies 2 is theorem 3 of [90]. The implication 2 implies 3 is theorem 1 of [49]. The equivalence of condition 4 is lemma 1 of [49]. The equivalence of conditions 5 and 6 to condition 2 is a special case of theorem 1 of [92]. We mention that argument. Conditions 5 and 6 are trivially equivalent to each other. Condition 2 implies condition 6 since the kernel of the character given in condition 2 is clearly a proper closed ideal containing both $(T - \lambda I)$ and $(T - \lambda I)^*$. The implication 6 implies 2 can be modeled on the second part of the proof of theorem 3 of [90], applied to the proper closed ideal given by condition 6.

William Arveson (theorem 3.1.2, page 181 of [5]) has shown that $\Lambda(T) \cap \partial W(T) = \Pi(T) \cap \partial W(T) \subset \Pi_n(T)$ where $\partial W(T)$ denotes the boundary of the numerical range $W(T)$ of T. An operator is said to be spectraloid if its spectral radius is equal to $\operatorname{Sup}\left\{|\mu| : \mu \in W(T)\right\}$. Thus it follows that $\Pi_n(T)$ is nonempty for any spectraloid operator T. (For another proof of this, see [84].)

If T is hyponormal (i.e., $TT^* \leq T^*T$) then $\Pi_n(T)$ is nonempty (cf. theorem 4 of [90] and [12]). Much more, John Bunce has shown that if T is hyponormal, then $\Pi_n(T) = \Pi(T)$, the ordinary approximate point spectrum of T (corollary 10 of [17]). Joe Stampfli has shown that all compact perturbations of hyponormal operators, as well as all compact perturbations of Toeplitz operators, have nonempty normal spectra [120]. This is based on the observation of Stampfli (easily deduced using the techniques found in the proof of proposition 3 of [79]) that the class of operators which have nonempty normal spectra is precisely the norm closure of the set of operators having one-dimensional reducing subspaces. This may be restated in a suggestive and possibly generalizable form as follows.

PROPOSITION 2.50. Every operator that weakly contains a one-dimensional operator is the norm limit of a sequence of operators which actually contain one-dimensional sub-operators.

In spite of the many positive results listed above it is not true that, in general, the normal spectrum $\Pi_n(T)$ is nonempty. Indeed Paul Halmos has shown (page 220 of [79]) that the set of operators having one-dimensional reducing spaces is not dense in $\mathcal{L}(\mathcal{H})$ ($\mathcal{H}$ an infinite dimensional separable Hilbert space) and this is equivalent to the fact that there exist operators whose one-dimensional part of $\hat{T}$ (i.e., $\Pi_n(T)$) is empty. (The example is simply the operator $\begin{pmatrix} 0 & 0 \\ I & 0 \end{pmatrix}$ on $\mathcal{H} \oplus \mathcal{H}$.) We next generalize this result by showing that there exists an operator (even irreducible) whose finite dimensional part of $\hat{T}$ is empty.

PROPOSITION 2.51. There exists an irreducible operator whose finite dimensional part of $\hat{T}$ is empty, i.e., every irreducible operator weakly contained in T is infinite dimensional.

PROOF. The example is the same operator we used to exhibit a non-smooth irreducible operator (proposition 1.53). The Topping operator is an operator T which generates a uniformly hyperfinite C^*-algebra (theorem and corollary 1 of [125]). Now suppose S is any operator weakly contained in T. Then there exists a non-trivial $*$-homomorphism φ of $C^*(T)$ onto $C^*(S)$ such that $\varphi(T) = S$. But the uniformly hyperfinite C^*-algebra $C^*(T)$ is simple by theorem 5.1, page 338 of [70] and hence φ is a $*$-isomorphism. Thus $C^*(S)$ is also a uniformly hyperfinite C^*-algebra and cannot act on a finite dimensional space. In fact $C^*(S)$ contains no nonzero compact operators. (Cf. also the remark following corollary 8 of [23].)

On the basis of the above proposition it is natural to conjecture that the set of operators on an infinite dimensional separable Hilbert space $\mathcal{H}$, with finite dimensional reducing subspaces, is not dense in $\mathcal{L}(\mathcal{H})$. As James Deddens has pointed out to the author, this conjecture is true and has been proven by James Williams in [128]. We have seen that the notion of an approximate normal proper value of an operator T is equivalent to the notion of a one-dimensional operator being weakly contained in T. One might hope for some analogue of this result, in which the notion of weak containment is characterized in spatial terms as a kind of "approximate containment." A recent attempt has been made in this direction by John Bunce and James Deddens [22] who have defined and studied a notion of one operator being a "subspace approximant" of another operator.

5. The topology on the spectrum

In the memoir we have adopted the point of view of George Mackey [96] in emphasizing, as the main structure on the spectrum and quasi-spectrum, the σ-ring of Borel sets. For that reason this section is independent and the reader primarily interested in the spectral multiplicity theory and unitary equivalence problem may proceed directly to the next chapter on decomposition theory.

Nevertheless it is true and fascinating that the weak containment concept (chapter 1, section 5) can be suitably extended to define a closure operation for sets which determines an interesting (in general non-Hausdorff) topology on the spectrum of an operator. In the case of smooth operators this topology is properly linked with the Borel structure already defined. So far at least, attempts at introducing a relevant topology in the quasi-spectrum have been abortive.

While the topological facts outlined in this section are not needed in the development to follow, the whole question of the relationship of properties of operators to topological properties of its spectrum is an intriguing and wide open research area. For example there is still another characterization of smooth operators as those for which the topology of its spectrum is T_o (for each pair of distinct points, there is a neighborhood of one to which the other does not belong). Further John Bunce and James Deddens have recently studied in detail the topology of the spectra of n-normal operators [20].

We have already indicated (cf. propositions 1.48 and 2.23) the basic connection between operators weakly contained in an operator T and $*$-representations of the C^*-algebra $C^*(T)$. (To each $*$-representation π of $C^*(T)$ one associates the operator $\pi(T)$.) By means of this connection essentially all of section 3 of [36] can be transplanted to the operator context, since $C^*(T)$ is a separable C^*-algebra with identity. For this reason we shall content ourselves with a basic description and properties of this topology and leave the details of the derivations to the interested reader. We shall basically adopt the point of view of Michael Fell [56], [57].

DEFINITION 2.52. Let T be an operator and let $\hat{T}$ be its spectrum. Let $\mathcal{S}$ be a subset of $\hat{T}$. We say an operator R is <u>weakly contained</u> in $\mathcal{S}$ if R is weakly contained in the operator $\Sigma \oplus S$ where the direct sum is taken of a set of concrete irreducible operators, one from each of the elements (unitary equivalence classes) of $\mathcal{S}$. Further an element $r \in \hat{T}$, is said to be weakly contained in $\mathcal{S}$, if any concrete operator R contained in r (a unitary equivalence class) is weakly contained in $\mathcal{S}$.

DEFINITION 2.53. If T is an operator and $\mathcal{S}$ is a subset of its spectrum $\hat{T}$, the closure of $\mathcal{S}$, denoted $\overline{\mathcal{S}}$, is defined to be the set of elements of $\hat{T}$ which are weakly contained in $\mathcal{S}$. A nontrivial verification shows that this closure operation satisfies the Kuratowski closure axioms (cf. page 43 of [91]) and hence defines a topology on $\hat{T}$.

The above topology, which might be called the Fell topology, or the weak containment topology is a very natural one and has various other descriptions. For example if $\hat{T}_n$ denotes that part of the spectrum $\hat{T}$ corresponding to n-dimensional irreducible operators, where n is a positive integer or $\aleph_0$, then the weak containment topology on $\hat{T}_n$ is precisely the same as the quotient topology obtained from the $*$-strong operator topology on the set of irreducible operators in $\mathcal{L}(\mathcal{H}_n)$. (The quotient is taken of course with respect to unitary equivalence.) (Cf. theorem 3.5.8 of [36] or theorem 3.1 of [56]. Cf. also [54], [66] and [67].)

Depending upon one's attitude the fact that this topology is not Hausdorff may be considered either its most pathological or its most fascinating feature. We take the latter view. Providing one is not fussy about Hausdorffness being a prerequisite for compactness, the spectrum $\hat{T}$ is compact, i.e., every open covering admits a finite subcovering (cf. proposition 3.1.8, page 61 of [36]). The topology of $\hat{T}$ does have some other redeeming qualities. For example it does have a countable base for its open sets (corollary to theorem 3.2, page 228 of [56]). Further $\hat{T}$ is a Baire space, i.e., if $\{V_i\}$ is a decreasing sequence of open sets, each one dense in $\hat{T}$, then $\cap V_i$ is also dense in $\hat{T}$ (corollary 3.4.13 of [36]).

One might conjecture that if T is an irreducible operator then its spectrum $\hat{T}$ would be a singleton set consisting of the unitary equivalence class of T. This is indeed not the case as one irreducible operator may have many other irreducible operators weakly contained in it. Of course for any irreducible operator T the class of T will be one element of $\hat{T}$ and it follows from the definition of the topology that that singleton set is in fact dense in $\hat{T}$. (Recall that in general $\hat{T}$ has rather poor separation properties and in general singleton sets are not closed.) Perhaps a specific example will help to picture this strange set of circumstances.

EXAMPLE 2.54. The spectrum $\hat{T}$ of the unilateral shift T is precisely $\tau \cup \{t\}$ where τ is the unit circle in $\mathbf{C}$ and t is the unitary equivalence class of T. The Fell topology on $\hat{T}$, when restricted to τ is the usual topology and the closure of $\{t\}$ is all of $\hat{T}$.

PROOF. Since the unilateral shift is hyponormal we know the one-dimensional part of $\hat{T}$ (i.e., the normal spectrum) is just the approximate point spectrum of T, which is well known to be the unit circle. Thus clearly $\tau \cup \{t\} \subset \hat{T}$ and it remains to show that there are no other (unitary equivalence classes of) irreducible operators weakly contained in T. Now it is easy to verify that the unilateral shift is a quasi normal operator (i.e., T commutes with T^*T) and thus that any operator weakly contained in T is also quasi normal. However it follows from the general structure theory for quasi normal operators (theorem 1 of [15]) that the only irreducible quasi normal operators (up to unitary equivalence) are one-dimensional (i.e., scalars) or positive scalar multiples of the unilateral shift itself. However the unilateral shift has the property that $T^*T = I$ and hence if π is any $*$-homomorphism of $C^*(T)$ one has $\pi(T^*T) = \pi(T)^*\pi(T) = I$. Thus if $\pi(T)$

is a scalar it must have absolute value 1 or if it is a positive scalar multiple of T that scalar multiple must be 1. Hence $\hat{T} \subset \tau \cup \{t\}$. The fact that the restriction of the topology of $\hat{T}$ to τ is the usual one just follows from the fact that the topology of the n-dimensional part of $\hat{T}$ is the quotient topology and, when $n = 1$, two scalars are unitary equivalent if and only if they are equal.

COROLLARY 2.55. The spectrum of the backward shift T^* is precisely $\tau \cup \{t^*\}$.

PROOF. Cf. proposition 2.42.

It follows from the work of Louis Coburn [27] [28] that every nonunitary isometry is weakly equivalent to the unilateral shift. Hence by proposition 2.35 above every nonunitary isometry S has this same spectrum $\hat{S} = \tau \cup \{t\}$.

Associated with any topology there is always a natural Borel structure (which we shall henceforth call the topological Borel structure) consisting of the σ-ring generated by the closed sets. If there is any justice we would expect this Borel structure to be the Mackey Borel structure already defined in $\hat{T}$ (definitions 2.26). As we shall see shortly justice only applies to the elite class of smooth operators. However, it is always true that the topological Borel structure is contained in the Mackey Borel structure (lemma 4.1 of [56]). In fact these topological considerations enable us to state still more characterizations of the class of smooth operators. See theorem 4.1 and lemma 4.2 of [56], theorem 1 of [41], as well as [69]. These facts are summarized in problem 9.5.6, page 185 of [36].

PROPOSITION 2.56. Let T be an operator. Then the following conditions are equivalent.

1. T is a smooth operator.
2. Given any two points in $\hat{T}$, there exists an open set which contains one and excludes the other.
3. The topological Borel structure for $\hat{T}$ makes $\hat{T}$ into a standard Borel space.
4. The Mackey Borel structure on $\hat{T}$ is identical to the topological Borel structure on $\hat{T}$.

We have already indicated that the Fell topology on the one-dimensional part of $\hat{T}$, i.e., the normal spectrum of T, is just the usual topology it has as a compact set of the complex plane. Actually quite a bit is known about the finite dimensional parts of $\hat{T}$. Let $\hat{T}_n$ denote that part of $\hat{T}$ corresponding to irreducible operators acting on a finite dimensional space of dimension n and let ${}_n\hat{T} = \bigcup_{m=1}^{n} \hat{T}_m$, where in each case n is a positive integer. Then ${}_n\hat{T}$ is a closed subset of $\hat{T}$ and $\hat{T}_n$ is a relatively open subset of ${}_n\hat{T}$ (proposition 3.6.3 page 74 of [36] and also [57]). For any finite integer n, the map $S \to \text{trace } S$ on $\hat{T}_n$ is continuous (proposition 3.6.4, page 74 of [36]). If for some particular n we have $\hat{T} = \hat{T}_n$, then the topology of $\hat{T}$ is Hausdorff (corollary 2, page 388 of [57] or theorem 4.2 of [88]). In general the norm function $S \to \|S\|$ on $\hat{T}$ is lower semi-continuous, i.e., if S_λ is a net in $\hat{T}$ converging to S in $\hat{T}$ then

$$\liminf \|S_\lambda\| \geq \|S\|$$

(cf. lemma 2.2 of [57] or proposition 3.3.2 of [36]). However if $\hat{T}$ happens to be Hausdorff, then the norm $S \to \|S\|$ is in fact continuous on $\hat{T}$ (corollary 3.3.9 of [36]). An interesting research problem would be the determination of the class of operators having Hausdorff-spectra. In fact John Bunce and James Deddens [24] have recently taken up just such a study. While they do not succeed in a complete characterization of those operators with Hausdorff spectra they do show it is a very restricted class of operators. For example they show that if $\hat{T}$ is Hausdorff then $\hat{T}$ must consist only of finite dimensional operators. If the elements of $\hat{T}$ are of bounded dimension (say bounded by the integer n) then T is called n-normal. Bunce and Deddens have found necessary and sufficient conditions for the spectrum of an n-normal operator to be Hausdorff. They have also shown that there exist operators with Hausdorff spectra which are not n-normal, for any n. (The reference to this author's work in [24] is to an earlier version of this memoir which went under a different title.)

We conclude this section on the topology of the spectrum with another example. One might conjecture that the spectrum of an irreducible compact operator is a singleton. If this were true we could of course use any infinite dimensional irreducible compact operator to illustrate the phenomenon of proposition 2.51. However the spectrum of an irreducible compact operator is a singleton if and only if that operator is finite dimensional. Indeed suppose T is an infinite dimensional irreducible operator. We will show that the assumption that $\hat{T}$ is a singleton leads to a contradiction. Indeed then $C^*(T)$, the C^*-algebra generated by T <u>and the identity</u> is a separable C^*-algebra with a one point spectrum, by corollary 2.31. But according to Rosenberg's theorem (cf. problem 4.7.3, page 96 of [36] this implies $C^*(T)$ is isomorphic to the algebra of all compact operators on some separable Hilbert space. Since $C^*(T)$ contains an identity it follows that the Hilbert space must be finite dimensional. Thus T weakly contains an irreducible finite dimensional operator. Thus the spectrum of T contains at least two points, a contradiction. In fact the spectrum $\hat{T}$ of an infinite dimensional irreducible compact operator T contains precisely <u>two</u> points, the unitary equivalence class τ of T and the one dimensional Null operator $\{0\}$. The closure of the singleton point τ is the two point set $\hat{T}$.

Chapter 3

DECOMPOSITION THEORY

1. The central decomposition

John von Neumann's idea of taking direct integrals of Hilbert spaces [127] (which he called generalized direct sums) is now some twenty-five years old and has led to many useful applications, particularly in the decomposition theory for rings of operators and in representation theory (both of locally compact groups and Banach algebras). For some reason, however, its great potential for operator theory has never been quite realized, although there has been some increased use of the theory in recent years (cf. section 2.3 of [6], as well as [68], [25] and [26]). However in representation theory even the choice of terminology (the "spectrum" of a C^*-algebra) indicates that von Neumann's reduction theory is in fact the natural extension of the spectral theorem for operators. This, of course, is precisely the thesis of this monograph.

We shall not, in this chapter, concern ourselves with all the various technical facts associated with various decompositions of operators but shall concern ourselves with precisely one (canonical) decomposition of an operator which is often called the central decomposition because it is modeled on part IV of von Neumann's reduction theory paper [127] where a ring of operators is decomposed with respect to its center. We shall show that this particular decomposition of an operator T into factor operators has many desirable properties (theorem 3.4 below) not the least of which is that the Borel space on which the measure is defined is precisely the quasi-spectrum $\tilde{T}$ of the operator and, when the operator T is normal, the decomposition reduces to the ordinary spectral theorem. Various simplifications occur when the operator T

is smooth and the situation there has been outlined in the first few pages of section 2.3 of [6].

We shall follow the presentation of von Neumann's direct integral theory as developed in chapter 2 of [37]. The need for notation that at times appears most imposing as well as the occasional digressions to deal with certain measure theoretic technicalities sometimes obscures the basic simplicity of the direct integral notion. The construction is just the natural generalization of the L^2 space of a measure space where one allows the square-integrable measurable functions to have values in a Hilbert space.

DEFINITION 3.1. For each fixed positive integer n, let $\mathcal{H}_n$ denote the Hilbert space of n-tuples of complex numbers and let $\mathcal{H}_0$ denote ℓ_2 (cf. definition 2.20). Then $\mathcal{L}$ will denote the space of operators $\mathcal{L} = \bigcup_{n=0}^{\infty} \mathcal{L}(\mathcal{H}_n)$. Let ν denote a positive Borel measure on a Borel space Z_n. Then we define a Hilbert space, denoting it by the following suggestive notation,

$$L^2(Z_n, \nu; \mathcal{H}_n)$$

as the set of all Borel maps $x : \xi \to x(\xi)$ of Z_n into $\mathcal{H}_n$ which are ν-square integrable in the sense that

$$\int_{Z_n} \|x(\xi)\|^2 \, d\nu(\xi) < +\infty,$$

and where we identify two such maps, x and y, if the L^2-norm of the difference is zero, i.e., if

$$\int_{Z_n} \|x(\xi) - y(\xi)\|^2 \, d\nu(\xi) = 0.$$

$L^2(Z_n, \nu; \mathcal{H}_n)$ is then a Hilbert space if one defines the linear structure and the inner product in the obvious manner.

This is the basic form of the direct integral concept that we shall need, as can be seen from the following considerations. Let ν now denote a positive Borel measure on a Borel space Z. Let $\xi \to \mathcal{H}(\xi)$ denote a Borel field of separable Hilbert spaces on Z (cf. definition 1, page 142 and remark 3, page 143 of [37]). Then by proposition 3, page 145 of [37] we have that the set

$$Z_n = \left\{\xi : \xi \in Z \text{ and } \dim \mathcal{H}(\xi) = n\right\}$$

is a Borel subset of Z and we thus obtain a partition of $Z = \bigcup_{n=0}^{\infty} Z_n$ and the direct integral of the Borel field $\xi \to \mathcal{H}(\xi)$ (definition 3 page 147 of [37]) is unitarily equivalent to an appropriate direct sum of generalized L^2-spaces, i.e.,

$$\int_Z^{\oplus} \mathcal{H}(\xi)\, d\nu(\xi) \simeq \sum_{n=0}^{\infty} \oplus L^2(Z_n, \nu; \mathcal{H}_n) .$$

We remark that a measure ν defined on a Borel space Z is called <u>standard</u> if there exists a ν-null set N in Z such that the Borel space $Z-N$ is a standard Borel space. If ν_n is standard then $L^2(Z_n, \nu; \mathcal{H}_n)$ is a separable Hilbert space (corollary, page 149 of [37]). Throughout this memoir we shall consider direct integral theory only with respect to standard measures. Further if ν_n and ν_n' are two equivalent positive measures on Z_n, then there is a canonical isomorphism of $L^2(Z_n, \nu_n; \mathcal{H}_n)$ onto $L^2(Z_n, \nu_n'; \mathcal{H}_n)$ which can be expressed in terms of the Radon-Nikodym derivative (remark, page 148 of [37].

A map of operators $\xi \to T(\xi)$ which is a Borel map of Z_n into $\mathcal{L}(\mathcal{H}_n)$ (cf. proposition 2.3) is ν-<u>essentially bounded</u> if the ν-essential supremum of the function $\xi \to \|T(\xi)\|$ is finite. Such a ν-essentially bounded Borel map of Z_n into $\mathcal{L}(\mathcal{H}_n)$ always determines a bounded linear operator T on $L^2(Z_n, \nu; \mathcal{H}_n)$ by defining

$$(Tx)(\xi) = T(\xi)x(\xi)$$

for all ξ in Z_n and all x in $L^2(Z_n, \nu; \mathcal{H}_n)$. The norm of T is of course the ν-ess. sup $\|T(\xi)\|$ and the operator T is denoted

$$T = \int_{Z_n}^{\oplus} T(\xi)d\nu(\xi)$$

(cf. definition 2, page 159 of [35]). More generally we use the notation

$$T = \int_{Z}^{\oplus} T(\xi)d\nu(\xi)$$

for the operator $\sum_{n=0}^{\infty} \oplus \int_{Z_n}^{\oplus} T(\xi)d\nu(\xi)$ acting on the space $\sum_{n=0}^{\infty} \oplus L^2(Z_n, \nu; \mathcal{H}_n)$ where $Z = \bigcup_n Z_n$ is a partition of Z.

Note that two ν-essentially bounded Borel maps of Z_n into $\mathcal{L}(\mathcal{H}_n)$ define the same operator on $L^2(Z_n, \nu; \mathcal{H}_n)$ if and only if they are equal ν almost everywhere (corollary, page 158 of [37]).

If f is a complex valued ν-essentially bounded Borel function on Z and I_n denotes the identity operator on the space $\mathcal{H}_n$, then $\xi \to f(\xi)I_n$ is a ν-essentially bounded Borel map of Z_n into $\mathcal{L}(\mathcal{H}_n)$, for $n = 0,1,2,\ldots$, and the corresponding operator

$$T_f = \sum_{n=0}^{\infty} \oplus \int_{Z_n}^{\oplus} f(\xi)I_n d\nu(\xi)$$

is called a <u>diagonalizable</u> <u>operator</u> on $\sum_{n=0}^{\infty} \oplus L^2(Z_n, \nu; \mathcal{H}_n)$.

An operator T acting on the space $\mathcal{H} = \sum_{n=0}^{\infty} \oplus L^2(Z_n, \nu; \mathcal{H}_n)$ is said to be <u>decomposable</u> if it is of the form

$$T = \sum_{n=0}^{\infty} \oplus \int_{Z_n}^{\oplus} T(\xi)d\nu(\xi)$$

where, for each n, $\xi \to T(\xi)$ is a ν - essentially bounded map of Z_n into $\mathcal{L}(\mathcal{H}_n)$. There is a very nice criterion for decomposability of an operator — namely an operator is decomposable on $\mathcal{H}$ if and only if it commutes with all the diagonalizable operators on $\mathcal{H}$ (corollary, page 164 of [37]).

In the material to follow we shall have occasion to refer also to the concept of direct integrals of von Neumann algebras and the relevant definitions and notation can be found in section 3 of chapter 2 of [37].

We now have summarized all the terminology and notation we need to describe the major fact about decomposition of operators, namely that there exists one canonical decomposition of any operator which has so many salutary qualities that it will be the only decomposition considered throughout this memoir. We first give a proposition which will serve as our definition of this (central) decomposition and then follow it by a theorem (3.4) which will extol its many desirable features. Finally one of its most agreeable properties, namely that the Borel space used in the decomposition can always be taken to be the quasi-spectrum of the operator, will be established (theorem 3.5).

PROPOSITION 3.2. Let S be an operator.

Part I (Existence). There exists an operator T unitarily equivalent to S, a Borel space Z with a Borel partition $Z = \bigcup_{n=0}^{\infty} Z_n$, a standard positive measure ν on Z such that T is a decomposable operator,

$$T = \int_{Z}^{\oplus} T(\xi)d\nu(\xi) = \sum_{n=0}^{\infty} \oplus \int_{Z_n}^{\oplus} T(\xi)d\nu(\xi)$$

on the space $\mathcal{H} = \sum_{n=0}^{\infty} \oplus L^2(Z_n, \nu; \mathcal{H}_n)$, where for each n, $\xi \to T(\xi)$ is a Borel map of Z_n into $\mathcal{L}(\mathcal{H}_n)$, and where the algebra of diagonalizable operators on $\mathcal{H}$ is precisely the center of $\mathcal{A}(T)$, the von Neumann algebra generated by T.

Part II (Uniqueness). Suppose T' is an operator unitarily equivalent to S which is decomposable with respect to some Borel space Z' and positive standard measure ν' on Z', i.e.,

$$T' = \int_{Z'}^{\oplus} T'(\xi)d\nu'(\xi)$$

such that the algebra of diagonalizable operators on the space

$$\mathcal{H}' = \sum_{n=0}^{\infty} \oplus L^2(Z_n', \nu'; \mathcal{H}_n)$$

is precisely the center of $\mathcal{A}(T')$. Then (after possibly eliminating a subset of Z of ν-measure 0 and a subset of Z' of ν'-measure 0) there exists a Borel isomorphism φ of Z onto Z' such that $\varphi(Z_n) = Z_n'$ and the measure ν'' induced on Z' by the map φ (i.e., $\nu''(B) = \nu(\varphi^{-1}(B))$ for every Borel set $B \subset Z'$) is equivalent to ν'. Thus there exists a natural linear isometry V of $\mathcal{H} = \sum_{n=0}^{\infty} \oplus L^2(Z_n, \nu; \mathcal{H}_n)$ onto $\mathcal{H}' = \sum_{n=0}^{\infty} \oplus L^2(Z_n', \nu'; \mathcal{H}_n)$. Further there exists an operator valued map $\xi \to U(\xi)$ on Z such that, when restricted to Z_n, it is a Borel map into the unitary operators in $\mathcal{L}(\mathcal{H}_n)$ and such that

$$U(\xi)T(\xi) = T'(\varphi(\xi))U(\xi)$$

for all $\xi \in Z$, and, if U denotes the unitary operator on

$\mathscr{H}$ defined by

$$U = \int_Z^{\oplus} U(\xi)d\nu(\xi) = \sum_{n=0}^{\infty} \oplus \int_{Z_n}^{\oplus} U(\xi)d\nu(\xi)$$

then

$$VUT = T'VU .$$

PROOF. Part I of this proposition is theorem 2 and its corollary, page 210 of [37] where the abelian von Neumann algebra referred to there is taken to be the center of $\mathcal{A}(T)$. Part II of the proposition is theorem 4, page 212 of [37].

DEFINITION 3.3. Proposition 3.2 states, with all the horrendous detail that preciseness imposes, that every operator T admits an essentially unique direct integral decomposition such that the algebra of diagonalizable operators may be identified with the center of $\mathcal{A}(T)$. Henceforth this decomposition will be called the <u>central decomposition of</u> T.

THEOREM 3.4. Let T be an operator with central decomposition

$$T = \int_Z^{\oplus} T(\xi)d\nu(\xi) = \sum_{n=0}^{\infty} \oplus \int_{Z_n}^{\oplus} T(\xi)d\nu(\xi)$$

on the space $\mathscr{H} = \sum_{n=0}^{\infty} \oplus L^2(Z_n, \nu; \mathscr{H}_n)$.

1. Each component operator $T(\xi)$ is a factor operator, for ν-almost all ξ in Z.
2. After possibly eliminating a set of ν-measure 0, the component operators are mutually disjoint, i.e., $T(\xi) \overset{|}{\circ} T(\eta)$ whenever ξ, η in Z, $\xi \neq \eta$.

3. The von Neumann algebra $\mathcal{A}(T)$ generated by T is a decomposable von Neumann algebra and in fact

$$\mathcal{A}(T) = \int_Z^{\oplus} \mathcal{A}(T(\xi))d\nu(\xi)$$

where $\mathcal{A}(T(\xi))$ of course denotes the factor von Neumann algebra generated by $T(\xi)$.

4. The commutant $\mathcal{A}(T)'$ of $\mathcal{A}(T)$ is a decomposable von Neumann algebra and in fact

$$\mathcal{A}(T)' = \int_Z^{\oplus} \mathcal{A}(T(\xi))' d\nu(\xi) .$$

5. The operator T is multiplicity free if and only if ν - almost all the component operators $T(\xi)$ are irreducible operators.

6. The operator T is finite (respectively semi-finite, infinite, of continuous type, type I, type II, type III) if and only if ν - almost all of the component operators $T(\xi)$ are finite (respectively semi-finite, infinite, of continuous type, type I, type II, type III).

7. If α is a complex number, then the central decomposition of αT is given by

$$\alpha T = \int_Z^{\oplus} \alpha T(\xi)d\nu(\xi) .$$

8. The central decomposition of T^* is given by

$$T^* = \int_Z^{\oplus} T(\xi)^* d\nu(\xi).$$

9. Except for a ν-null subset of Z, all the component operators $T(\xi)$ are weakly contained in T.

PROOF. Properties 1 and 3 are contained in theorem 2 and its corollary, page 210 of [37]. Applying theorem 4, page 176, to property 3 gives property 4. To obtain property 5 apply the corollary to proposition 6, page 181 of [37] to the decomposition of $\mathcal{A}(T)'$ given by property 4, where the projection E in that corollary is taken to be the identity in $\mathcal{A}(T)'$. Thus T is multiplicity free (i.e., $\mathcal{A}(T)'$ is abelian) if and only if ν-almost all the component von Neumann algebras $\mathcal{A}(T(\xi))'$ are abelian. But since we also know by property 1 that ν-almost all the components $\mathcal{A}(T(\xi))'$ are factors we have that $\mathcal{A}(T(\xi))'$ consists only of complex multiples of the identity operator for ν-almost all ξ. Thus $T(\xi)$ is irreducible for ν-almost all ξ. The argument is reversible. By property 4 and proposition 1.38 the verification of property 6 is reduced to the corresponding statements for von Neumann algebras. Thus the assertion for finite and infinite operators follows from corollary 2, page 207 of [37]. The assertion for discrete (i.e., type I) operators and operators of continuous type follows from corollary 2, page 183 of [37]. The result for semi-finite operators and type III (i.e., purely infinite) operators is obtained from corollary 2, page 206 of [37]. The type II case of course follows from these previous results since type II operators are defined to be semi-finite operators of continuous type.

Properties 7 and 8 follow from proposition 3, page 159 of [37], together with the easy verifications that the decompositions are indeed central.

To establish property 2 we must give a slightly modified version of the proof of proposition 3 of [50]. The basic ideas involved in this proof are due to Alain Guichardet [73]. Since T generates $\mathcal{A}(T)$ if S is any operator in $\mathcal{Q}(T)$, there exists a sequence $\{p_i\}$ of polynomials in two non-commuting indeterminants (cf. definition 1.31) such that $p_i(T,T^*)$ converges strongly to S. Since every operator in $\mathcal{Q}(T)$ is decomposable (property 3) we have

$$S = \int_Z^{\oplus} S(\xi)d\nu(\xi)$$

where $\xi \to S(\xi) \in \mathcal{Q}(\xi) \subset \mathcal{L}(\mathcal{H}_n)$ is a Borel map on Z_n, for $n = 0,1,2,\ldots$. Also by proposition 3, page 159 of [37] we have

$$p_i(T,T^*) = \int_Z^{\oplus} p_i\left(T(\xi),\, T(\xi)^*\right)d\nu(\xi)$$

for $i = 1,2,\ldots$.

Since $p_i(T,T^*)$ converges to S strongly, proposition 4, part i), page 160 of [37] implies that, after possibly taking a subsequence of $\{p_i\}$, and eliminating a set of ν-measure zero from Z, we may assume without loss of generality that $p_i\left(T(\xi),\, T(\xi)^*\right)$ converges strongly to $S(\xi)$ for all ξ in Z.

Now suppose ξ and η are two points of Z and $T(\xi)$ and $T(\eta)$ are not disjoint. Then there exists a nonzero operator $R(\xi, \eta)$ which inter-twines $T(\xi)$ and $T(\eta)$, i.e.,

$$R(\xi, \eta)\, T(\xi) = T(\eta)\, R(\xi, \eta)$$

and

$$R(\xi, \eta)\, T(\xi)^* = T(\eta)^* R(\xi, \eta)$$

(cf. definition 1.3 and corollary 1.8). Then clearly $R(\xi, \eta)$ will inter-twine each $p_i\left(T(\xi),\, T(\xi)^*\right)$ and hence by strong-convergence will intertwine $S(\xi)$ and $S(\eta)$.

Since ν is standard we may assume without loss of generality that Z

is a standard Borel space. Let $\{B_n\}$ denote a countable separating family of Borel sets of Z. To each such set B_n we may associate the diagonalizable operator E_n, belonging to the characteristic function of the set B_n. Then each E_n is a projection in the center of $\mathcal{A}(T)$. Applying our previous result for an arbitrary operator S in $\mathcal{A}(T)$, to each of the operators E_n in $\mathcal{A}(T)$ (and in the process perhaps eliminating a countable number of null sets from Z) we have that,

$$R(\xi, \eta)E_n(\xi) = E_n(\eta)R(\xi, \eta)$$

for all n. But by definition $E_n(\xi)$ is the identity operator if $\xi \in B_n$ and is otherwise the zero operator. Since $R(\xi, \eta)$ is nonzero we have $E_n(\xi) = E_n(\eta)$ for all n. Since the family $\{B_n\}$ separates points we have $\xi = \eta$.

Finally we prove property 9. By eliminating a ν-null set we may assume that $\|T(\xi)\| \leq \|T\|$ for all ξ in Z. Let $Q(T)$ denote the set of operators in $C^*(T)$ of the form $p(T,T^*)$ where p is a polynomial in two non-commuting indeterminates, and the scalar coefficients are of the form $\alpha + \beta i$ where α and β are rational numbers and $i = \sqrt{-1}$. Then $Q(T)$ is a countable set norm dense in $C^*(T)$. Further, by proposition 3, page 159 of [37] we have that each $p(T,T^*)$ is a decomposable operator and furthermore

$$p(T,T^*) = \int_Z^{\oplus} p\left(T(\xi), T(\xi)^*\right) d\nu(\xi).$$

Hence

$$\|p(T,T^*)\| \leq \nu\text{-ess. sup}\, \| p\left(T(\xi), T(\xi)^*\right) \|.$$

Thus for each p in $Q(T)$, there exists a null-set N_p such that

$$\|p(T(\xi), T(\xi)^*)\| \le \|p(T,T^*)\|$$

for all ξ in Z, $\xi \notin N_p$. Since $Q(T)$ is a countable set we may eliminate from Z a ν-null set and assume, without loss of generality, that

$$\|p(T(\xi), T(\xi)^*)\| \le \|p(T,T^*)\|$$

for all ξ in Z and all $p \in Q(T)$. For each ξ in Z, define a map π_ξ of $Q(T)$ into $C^*(T(\xi))$ by

$$\pi_\xi\left(p(T,T^*)\right) = p\left(T(\xi), T(\xi)^*\right).$$

Then π_ξ is a norm continuous $*$-homomorphism of $Q(T)$ into $C^*(T(\xi))$ such that $\pi_\xi(T) = T(\xi)$. Since $Q(T)$ is norm dense in $C^*(T)$, π_ξ has a unique extension to a continuous $*$-homomorphism of $C^*(T)$ into $C^*(T(\xi))$, which maps T into $T(\xi)$ and hence must be surjective. Thus for each ξ, $T(\xi)$ is weakly contained in T.

Our next result corresponds to theorem 2 of [50].

THEOREM 3.5. The central decomposition of any operator T may always be taken over its quasi-spectrum $\tilde{T}$. More explicitly, there exists a standard finite Borel measure ν on $\tilde{T}$ and a Borel map $\xi \to T(\xi)$ of $\tilde{T}$ into $\mathcal{L} = \bigcup_{n=0}^{\infty} \mathcal{L}(\mathcal{H}_n)$ such that the operator $T(\xi)$ is an element of the quasi-equivalence class ξ, for ν-almost all ξ and

$$T \simeq \int_{\tilde{T}}^{\oplus} T(\xi)\,d\nu(\xi)$$

is the central decomposition of T.

PROOF. Let

$$T \simeq \int_Z^{\oplus} T(\xi)d\nu(\xi)$$

denote the central decomposition of T, where Z is a standard Borel space, ν a positive finite measure on Z and $\xi \to T(\xi)$ is a Borel map of Z into $\mathcal{L} = \bigcup_{n=0}^{\infty} \mathcal{L}(\mathcal{H}_n)$. By parts 1, 2 and 9 of theorem 3.4 we may assume, without loss of generality, that the components $T(\xi)$ are mutually disjoint factor operators weakly contained in T. Thus the map $\psi: \xi \to T(\xi)$ is a one-to-one Borel map of Z into the standard Borel space T^f, the subset of the concrete dual of T consisting of factor operators (cf. definition 2.24 and corollary 2.25). Hence by theorem 3.2 of [96] the range $\psi(Z)$ of ψ is a standard Borel subspace of T^f and ψ is a Borel isomorphism of Z onto $\psi(Z)$. Let φ denote the canonical (quotient) mapping of T^f onto $\tilde{T}$ defined by sending a factor operator in T^f, into the quasi-equivalence class containing it. Since the elements of $\psi(Z)$ are two-by-two disjoint, the map φ, restricted to $\psi(Z)$ is a one-to-one Borel map of $\psi(Z)$ into $\tilde{T}$. Since $\psi(Z)$ intersects each quasi-equivalence class in at most one point, corollary 2.17 implies that $[\psi(Z)]^q$, the subset of T^f consisting of all operators quasi-equivalent to any element of $\psi(Z)$, is a Borel subset of T^f. Hence by the definition of the quotient Borel structure, the canonical image $\varphi(\psi(Z))$ in $\tilde{T}$ is a Borel subset of $\tilde{T}$.

Since φ is a one-to-one Borel map on $\psi(Z)$, to show it is a Borel isomorphism we need only show that it maps Borel sets into Borel sets. Let B be a Borel subset of $\psi(Z)$. Then B^q is a Borel subset of T^f, again by corollary 2.17, and hence $\varphi(B)$ is a Borel subset of $\tilde{T}$ and hence of $\varphi(\psi(z))$.

Hence we have established that

$$\varphi \circ \psi : \xi \to \varphi(T(\xi))$$

is a Borel isomorphism of the standard space Z onto the standard space $\varphi(\psi(Z)) \subset \tilde{T}$. Thus we may transfer the measure ν and the map $\xi \to T(\xi)$ from Z to $\varphi(\psi(Z))$ and then extend the domain of definition of the resulting measure to all of $\tilde{T}$ by defining the relative complement of $\varphi(\psi(Z))$ in $\tilde{T}$ to be a null-set. The result now follows immediately.

DEFINITION 3.6. Note that, by part II of proposition 3.2, the previous theorem associates, with each operator T, a unique standard finite measure class ν on its quasi-spectrum $\tilde{T}$, since in any case the measure is determined only up to equivalence. We call this measure class on $\tilde{T}$ the *central measure class* associated with T.

REMARK 3.7. We are now ready to add a refinement to our charting procedure. Up to now we have specified the location of an operator by only one "coordinate," namely its quasi-spectrum. Proposition 2.35 states that two operators are weakly equivalent if and only if they have the same quasi-spectrum. The mesh of this chart is very broad and we next wish to introduce a second "coordinate" which will help to locate operators within a weak equivalence class. This second "coordinate" (a measure class on the quasi-spectrum) will enable us to distinguish operators up to quasi-equivalence. This result (theorem 3.8) is an operator theoretic analogue of theorem 3 of [50]. Unfortunately, under this charting procedure, not every measure class on the quasi-spectrum need correspond to a quasi-equivalence class of operators. The formidable task of identifying those measure classes (the central measure classes) which do, will be taken up immediately following theorem 3.8. Then, in the next chapter, we will further refine the mesh of our chart by specifying a third "coordinate," a spectral multiplicity function, which will enable one to distinguish operators, up to unitary equivalence, within a quasi-equivalence class.

THEOREM 3.8. Two operators S and T are quasi-equivalent ($S \approx T$) if and only if they have the same quasi-spectrum and the same central measure class on that quasi-spectrum.

PROOF. Suppose first that $T \approx S$ and that

$$T \simeq \int_{\tilde{T}}^{\oplus} T(\xi)d\nu(\xi)$$

is the central decomposition of T. For each n, $n = 0,1,2,\ldots$, let U_n denote a linear isometry of $\textcircled{\infty}\mathcal{H}_n$ (the direct sum of a countable number of copies of $\mathcal{H}_n$) onto $\mathcal{H}_0 = \ell_2$. Then the map

$$\xi \to U_n \textcircled{\infty} T(\xi)\, U_n^{-1}$$

of $\tilde{T}(n)$ into $\mathcal{L}(\mathcal{H}_0) \subset \mathcal{L} = \bigcup_{n=0}^{\infty} \mathcal{L}(\mathcal{H}_n)$ is a Borel map on $\tilde{T}(n)$ where $\tilde{T} = \bigcup_{n=0}^{\infty} \tilde{T}(n)$ is a Borel partition of $\tilde{T}$ where

$$\tilde{T}(n) = \left\{\xi : \xi \in \tilde{T} \text{ and } T(\xi) \in \mathcal{L}(\mathcal{H}_n)\right\} .$$

Then one can verify that

$$\textcircled{\infty} T \simeq \int_{\tilde{T}}^{\oplus} U_n \textcircled{\infty} T(\xi)\, U_n^{-1}\, d\nu(\xi)$$

is the central decomposition of $\textcircled{\infty}T$. (Since $\textcircled{\infty}T$ and T are weakly equivalent they have the same quasi-spectrum $\tilde{T}$ by proposition 2.35. Further the center of $\mathcal{A}(\textcircled{\infty}T)$ is just $\left\{\textcircled{\infty}R : R \in \mathcal{A}(T) \cap \mathcal{A}(T)'\right\}$.) Thus both T and $\textcircled{\infty}T$ determine the same central measure class on $\tilde{T}$. Similarly S and $\textcircled{\infty}S$ determine the same central measure class on $\tilde{S} = \tilde{T}$. But since S and T are quasi-equivalent, $\textcircled{\infty}S$ and $\textcircled{\infty}T$ are unitarily equivalent by proposition 1.11. It follows that S and T determine the same central measure class on $\tilde{T}$.

Conversely suppose S and T have the same quasi-spectrum $\tilde{T}$ and the same central measure class (with representative) ν. Then S and T admit central decompositions of the form

$$S \simeq \int_{\tilde{T}}^{\oplus} S(\xi)d\nu(\xi) \quad \text{and} \quad T \simeq \int_{\tilde{T}}^{\oplus} T(\xi)d\nu(\xi).$$

By identifying two unitarily equivalent operators, we may assume, without loss of generality, that the two unitary equivalences above are in fact equalities. This agreement will ensure that the argument is not obscured by a proliferation of unessential linear isometries. Since $S(\xi)$ and $T(\xi)$ are both in the same quasi-equivalence class (namely ξ) we have $S(\xi) \approx T(\xi)$ for all $\xi \in \tilde{T}$. Thus by proposition 1.34, for each $\xi \in \tilde{T}$ there exists a $*$-isomorphism φ_ξ of $\mathcal{A}(T(\xi))$ onto $\mathcal{A}(S(\xi))$ such that $\varphi_\xi(T(\xi)) = S(\xi)$. We claim that the map $\xi \to \varphi_\xi$ is a ν-measurable field of isomorphisms, in the sense of section 3.6 of chapter 2 of [37], i.e., if $\xi \to A(\xi) \in \mathcal{A}(T(\xi))$ is a Borel field of operators on $\tilde{T}$, then the map $\xi \to \varphi_\xi(A(\xi)) \in \mathcal{A}(S(\xi))$ is a Borel field of operators on B, where B is a Borel subset of $\tilde{T}$ whose relative complement is a ν-null set.

If $\xi \to A(\xi) \in \mathcal{A}(T(\xi))$ is a Borel field of operators on $\tilde{T}$ then

$$A = \int_{\tilde{T}}^{\oplus} A(\xi)d\nu(\xi) = \int_{B}^{\oplus} A(\xi)d\nu(\xi) \in \mathcal{A}(T)$$

since

$$\mathcal{A}(T) = \int_{\tilde{T}}^{\oplus} \mathcal{A}(T(\xi))d\nu(\xi)$$

by part 3 of theorem 3.4. Thus there exists a sequence of polynomials $\{p_i\}$ in two, noncommuting indeterminates such that the sequence of operators

$\{p_i(T,T^*)\}$ converges strongly to A. By proposition 3, page 159 of [37] we have

$$p_i(T,T^*) = \int_B^{\oplus} p_i\left(T(\xi), T(\xi)^*\right) d\nu(\xi).$$

By taking a subsequence of the original sequence $\{p_i\}$, and eliminating another set of ν-measure zero we may assume, without loss of generality, that $p_i\left(T(\xi), T(\xi)^*\right)$ converges strongly to $A(\xi)$ for all ξ in B (cf. proposition 4, page 160 of [37]).

However each of the isomorphisms φ_ξ have the property that

$$\varphi_\xi\left(p_i(T(\xi), T(\xi)^*)\right) = p_i\left(S(\xi), S(\xi)^*\right)$$

and hence $p_i\left(S(\xi), S(\xi)^*\right)$ converges strongly to $\varphi_\xi(A(\xi))$ for all ξ in B. Since each of the maps

$$\xi \to p_i\left(S(\xi), S(\xi)^*\right)$$

is a Borel map on B we conclude that the map $\xi \to \varphi_\xi(A(\xi))$ is a Borel map on B. We have thus verified that $\xi \to \varphi_\xi$ is a ν-measurable field of isomorphisms. Hence by proposition 10, page 185 of [37] we have that

$$\varphi(A) = \int_B^{\oplus} \varphi_\xi(A(\xi))d\nu(\xi)$$

defines an isomorphism of $\mathcal{A}(T)$ onto $\mathcal{A}(S)$ such that $\varphi(T) = S$. By proposition 1.34 we conclude that S and T are quasi-equivalent.

2. The central measure lattice

DEFINITION 3.9. Recall that $\tilde{\mathfrak{F}}$ denotes the (non-smooth) Borel space of quasi-equivalence classes of factor operators in $\mathcal{L} = \bigcup_{n=0}^{\infty} \mathcal{L}(\mathcal{H}_n)$ (cf. definitions 2.20 and 2.26). Every operator T determines a measure class on

$\tilde{\mathfrak{J}}$ simply by taking the central measure class determined by T on the Borel subset $\tilde{T}$ of $\tilde{\mathfrak{J}}$ and extending its domain of definition to all of $\tilde{\mathfrak{J}}$ by requiring that the complement of $\tilde{T}$ in $\tilde{\mathfrak{J}}$ be a null-set. The collection of all such central measure classes defined on $\tilde{\mathfrak{J}}$ and determined by operators in $\mathcal{L}$ will be called the central measure lattice on $\tilde{\mathfrak{J}}$ and will be denoted $C\mathfrak{M}(\mathfrak{J})$. Each such central measure on $\tilde{\mathfrak{J}}$ is in fact determined only by the quasi-equivalence class of the operator T in $\mathcal{L}$ (theorem 3.8) and hence there is a natural map of the space $\tilde{\mathcal{L}}$ of quasi-equivalence classes of operators in $\mathcal{L}$, onto the central measure lattice $C\mathfrak{M}(\tilde{\mathfrak{J}})$. The sole purpose of this admittedly rather technical section is the identification of that central measure lattice.

PROPOSITION 3.10. The natural map of $\tilde{\mathcal{L}}$ onto $C\mathfrak{M}(\tilde{\mathfrak{J}})$ is a lattice isomorphism. Specifically if s, t are elements of $\tilde{\mathcal{L}}$ then s covers t $(s \,\}\, t)$ if and only if $C(s) >> C(t)$, where $C(s)$ and $C(t)$ denote the standard measure classes on $\tilde{\mathfrak{J}}$ corresponding to s and t respectively.

PROOF. The natural map is one-to-one by theorem 3.8. Suppose first that $s \,\}\, t$ and that S (respectively T) is an operator from $\mathcal{L}$ which is contained in the quasi-equivalence class s (respectively t). Then $S \,\}\, T$ and hence, by proposition 2.12 and proposition 1.35, part 3, we may write $S = S_1 \oplus S_2$ such that $S_1 \stackrel{|}{\circ} S_2$, $S_2 \stackrel{|}{\circ} T$ and $T \,\}\, S_1$ and hence $S_1 \approx T$. Let E denote the central projection of $\mathcal{Q}(S)$ such that S_1 is the restriction of S to the range of E. Let μ be a representative standard finite positive measure from the class $C(s)$ and let

$$S \simeq \int_{\tilde{S}}^{\oplus} S(\xi)\, d\mu(\xi)$$

denote the central decomposition of S. Thus the central projection E corresponds to a diagonalizable operator and since it is a projection it in fact corresponds to a characteristic function of a Borel subset $B \subset \tilde{S}$ and further

$$S_1 \simeq \int_B^{\oplus} S(\xi)\, d\mu(\xi).$$

Define the measure ν on $\tilde{S}$ by $\nu(C) = \mu(C \cap B)$ for all Borel sets C contained in $\tilde{S}$. Then S_1 has central decomposition

$$S_1 \simeq \int_{\tilde{S}}^{\oplus} S(\xi)\, d\nu(\xi)$$

and clearly $\mu >> \nu$. However since $S_1 \approx T$ we have (by theorem 3.8) that ν is contained in the measure class $\mathcal{C}(t)$ corresponding to T. Thus $\mathcal{C}(s) >> \mathcal{C}(t)$.

Conversely, suppose $\mathcal{C}(s) >> \mathcal{C}(t)$ and μ and ν are representatives of each of the classes $\mathcal{C}(s)$ and $\mathcal{C}(t)$ respectively. Then we may write $\mu = \mu_1 \vee \mu_2$ where $\mu_1 \simeq \nu$, $\mu_2 \perp \nu$ and $\mu_1 \perp \mu_2$. Indeed there exists a Borel set $B \subset \tilde{\mathfrak{F}}$ such that $\mu_1(C) = \mu(C \cap B)$ and $\mu_2(C) = \mu(C \cap (\tilde{\mathfrak{F}} - B))$ for all Borel sets $C \subset \tilde{\mathfrak{F}}$ (cf. theorem 2, page 78 of [77]). Suppose S (respectively T) is an operator from $\mathcal{L}$ which is contained in the quasi-equivalence class s (respectively t). Taking the central decomposition of S with respect to the measure μ gives

$$S \simeq \int_{\tilde{\mathfrak{F}}}^{\oplus} S(\xi)\, d\mu(\xi)$$

$$\simeq \int_{\tilde{\mathfrak{F}}}^{\oplus} S(\xi)\, d\mu_1(\xi) \oplus \int_{\tilde{\mathfrak{F}}}^{\oplus} S(\xi)\, d\mu_2(\xi).$$

Further $\int_{\tilde{\mathfrak{F}}}^{\oplus} S(\xi)\, d\mu_1(\xi)$ is the central decomposition of S_1 and hence the fact that $\mu_1 \simeq \nu$ implies that $S_1 \approx T$ by theorem 3.8. Hence $S \gtrsim S_1 \approx T$ or $S \} T$. Thus $s \} t$.

COROLLARY 3.11. The collection $C\mathcal{M}(\tilde{\mathfrak{F}})$ of central measure classes on $\tilde{\mathfrak{F}}$ is a lattice ideal of the lattice $\mathcal{M}(\tilde{\mathfrak{F}})$ of all standard finite positive measure classes on $\tilde{\mathfrak{F}}$.

PROOF. Let $\tilde{\mu} \in C\mathcal{M}(\tilde{\mathfrak{F}})$ and $\tilde{\nu} \in \mathcal{M}(\tilde{\mathfrak{F}})$ and $\tilde{\nu} << \tilde{\mu}$. By $C\mathcal{M}(\tilde{\mathfrak{F}})$ being a lattice ideal we mean that these conditions imply that $\tilde{\nu} \in C\mathcal{M}(\tilde{\mathfrak{F}})$, which fact we now prove. Let μ (respectively ν) denote a member of the class $\tilde{\mu}$ (respectively $\tilde{\nu}$). Then there exists a Borel set $B \subset \tilde{\mathfrak{F}}$ such that the measure μ_B defined on $\tilde{\mathfrak{F}}$ by $\mu_B(C) = \mu(B \cap C)$ for all Borel subsets C of $\tilde{\mathfrak{F}}$, is equivalent to ν (cf. theorem 2, page 78 of [77]). Since $\tilde{\mu} \in C\mathcal{M}(\tilde{\mathfrak{F}})$ there exists an operator S in $\mathcal{L}$ with central decomposition

$$S \simeq \int_{\tilde{\mathfrak{F}}}^{\oplus} S(\xi)\, d\mu(\xi).$$

Then

$$T = \int_B^{\oplus} S(\xi)\, d\mu(\xi) = \int_{\tilde{\mathfrak{F}}}^{\oplus} S(\xi)\, d\mu_B(\xi)$$

$$\simeq \int_{\tilde{\mathfrak{F}}}^{\oplus} S(\xi)\, d\nu(\xi)$$

is a central decomposition. Thus $\nu \in C\mathcal{M}(\tilde{\mathfrak{F}})$.

COROLLARY 3.12. If $s,t \in \tilde{\mathcal{L}}$ and $C(s)$ and $C(t)$ denote the corresponding central measure classes on $\tilde{\mathfrak{F}}$, then $s \between t$ if and only if $C(s) \perp C(t)$.

Note that two quasi-equivalent operators necessarily have the same norm (cf. proposition 1.34). Thus we may, for any $t \in \tilde{\mathcal{L}}$, denote by $\|t\|$ the common norm of the operators in the class t. Of course this does not define a norm on $\tilde{\mathcal{L}}$ since $\tilde{\mathcal{L}}$ does not have a linear structure. Thus we will denote $\tilde{\mathcal{L}}_r = \{t : t \in \tilde{\mathcal{L}} \text{ and } \|t\| \leq r\}$, for any $r > 0$. Similarly $\tilde{\mathfrak{I}}_r = \tilde{\mathfrak{I}} \cap \tilde{\mathcal{L}}_r$. We also say a measure μ on $\tilde{\mathfrak{I}}$ has support bounded by r if the relative complement of $\tilde{\mathfrak{I}}_r$ is a μ - null set. Thus under the lattice isomorphism given by proposition 3.10, $\tilde{\mathcal{L}}_r$ corresponds to those measure classes in $C\mathcal{M}(\tilde{\mathfrak{I}})$ which have support bounded by r (cf. proposition 1.58).

It is clear that $C\mathcal{M}(\tilde{\mathfrak{I}})$ is a proper ideal of $\mathcal{M}(\tilde{\mathfrak{I}})$ since every element $C(t)$ of $C\mathcal{M}(\tilde{\mathfrak{I}})$ has bounded support (by $\|t\|$). While one might expect that this condition of bounded support would be sufficient to characterize the central measure lattice it is one of the nasty surprises of the theory that this is not the case. Thus there exists a finite positive standard measure with bounded support, on $\tilde{\mathfrak{I}}$, which is not a central measure, i.e., is not of the form $C(t)$ for any $t \in \tilde{\mathcal{L}}$. Indeed let T be any non-smooth operator. (Such operators exist by proposition 1.53.) Then $\tilde{T}$ is a bounded Borel subset of $\tilde{\mathfrak{I}}$, $\tilde{T} \subset \tilde{\mathfrak{I}}_{\|T\|}$ and the (non-smooth) Borel space $\tilde{T}$ admits a finite positive standard measure which is not central. Indeed by corollary 2.31 we may identify $\tilde{T}$ with $C^*(T)^\sim$, the quasi-spectrum of $C^*(T)$, and it is not difficult to show that a central measure on $\tilde{T}$ corresponds to the central decomposition of some operator S weakly contained in T (cf. corollary 2.38) and thus corresponds to a central measure on $C^*(T)^\sim$, i.e., a measure associated with the central decomposition of a $*$ - representation of $C^*(T)$. But Jacques Dixmier has shown (cf. corollary 4, page 4186 of [39]) that whenever T is not smooth, i.e., $C^*(T)$ is not GCR, that $C^*(T)^\sim$ admits finite positive standard measures which do not arise from central decompositions. Thus such non-central standard measures exist on $\tilde{T}$ whenever $\tilde{T}$ is not smooth, and clearly their extension to $\tilde{\mathfrak{I}}$ have support bounded by

$\|T\|$. Thus we must continue the technical task of identifying the central measures on $\tilde{\mathfrak{T}}$. However we have still one more characterization of smooth operators that says this problem just doesn't exist if one wishes to use measures on $\tilde{T}$ to classify, up to quasi-equivalence, all the operators weakly contained in some smooth operator T.

PROPOSITION 3.13. An operator T is smooth if and only if every standard finite positive measure class on $\tilde{T}$ is central.

PROOF. We have just noted that $\tilde{T}$ has noncentral standard measures whenever T is not smooth. The positive statement for smooth operators follows also from known results for GCR C^*-algebras (cf. proposition 4.61, remark 7.3.7 and theorem 8.6.5 of [36]) and the identification of $\tilde{T}$ with $C^*(T)^\sim$.

We now proceed to try to identify the central measure classes on $\tilde{T}$ even when T is not smooth and hence finally to identify the central measure classes on $\tilde{\mathfrak{T}}$. There is a rather unsatisfactory characterization of the central measures on $C^*(T)^\sim$ (and hence on $\tilde{T}$) given by proposition 7 of [50]. Further Edward Effros has a method of obtaining central measures on $C^*(T)^\sim$ (and hence on $\tilde{T}$) (theorem 4.3 of [45]) but unfortunately that procedure does not actually characterize the central measures. Also Alain Guichardet has obtained a number of interesting sufficient conditions for a measure class to be central [75]. More recently Effros has managed to characterize the central measures on the quasi-spectrum of a separable C^* - algebra [48] based on the work of Brian Davies [31], [32], [33]. It is this technique of Effros that we now apply to obtain our main characterization (theorem 3.18

below) of the central measure lattice $\mathcal{C}\mathcal{M}(\tilde{\mathfrak{F}})$ on $\tilde{\mathfrak{F}}$. We mention that what we have called central measures, Effros refers to as canonical measures in [48].

DEFINITION 3.14. We define the <u>big operator of norm r</u> to be the operator

$$A(r) = \sum_{T \in \mathfrak{F}_r} \oplus T$$

where as usual $\mathfrak{F}_r$ denotes the set of factor operators in $\mathcal{L}$ of norm less than or equal to r. Clearly $\|A(r)\| = r$ and $A(r)$ acts on an enormous (nonseparable) Hilbert space.

For any operator T, let $\Sigma^*(T)$ denote the smallest C^*-algebra of operators containing T and the identity operator and which is also closed under weak <u>sequential</u> convergence. If T acts on a separable Hilbert space Richard Kadison has shown that $\Sigma^*(T)$ is just the ordinary von Neumann algebra $\mathcal{A}(T)$ generated by T (theorem A of the appendix to [31]). However in general $\Sigma^*(T)$ may be considerably smaller than $\mathcal{A}(T)$, in which case $\Sigma^*(T)$ is a Σ^*-algebra defined and studied by Brian Davies (cf. page 149 of [31] and also [33]). While generally we are concerned only with operators acting on separable spaces, we shall need this concept as we shall apply it to the big operator of norm r which acts on a nonseparable Hilbert space.

DEFINITION 3.15. A subset $B \subset \tilde{\mathfrak{F}}_r$ is called <u>Davies-Borel</u> if the projection

$$E_B = \sum_{T \in \mathfrak{F}_r} \oplus \chi_B(T^q)\, I_T$$

is contained in $\Sigma^*(A(r))$, where χ_B is the characteristic function of B,

T^q is the quasi-equivalence class containing T, I_T denotes the identity operator on the space $\mathcal{H}(T)$ of T, and $A(r)$ denotes the big operator of norm r.

Clearly every operator of the form E_B commutes with $A(r)$ and hence the condition of definition 3.15 may be equivalently formulated as requiring that the projection E_B be contained in the center of $\Sigma^*(A(r))$.

PROPOSITION 3.16. The collection of Davies-Borel sets in $\tilde{\mathfrak{J}}_r$ forms a Borel structure for $\tilde{\mathfrak{J}}_r$ which is contained in the Mackey Borel structure of $\tilde{\mathfrak{J}}_r$ (cf. definitions 2.26). Furthermore $\tilde{\mathfrak{J}}_r$ is the quasi-spectrum of $A(r)$ which may be identified with the quasi-spectrum of the separable C^*-algebra, $C^*(A(r))$ in the usual way (corollary 2.31). Under this identification the Davies-Borel structure defined on $\tilde{\mathfrak{J}}_r$ corresponds to the Davies-Borel structure defined on the quasi-spectrum of $C^*(A(r))$ as defined by Brian Davies (cf. theorem 4.2 of [32]).

PROOF. Clearly $\tilde{\mathfrak{J}}_r$ is the quasi-spectrum of $A(r)$ since every element of $\mathfrak{J}_r$ is actually contained and hence weakly contained in the big operator $A(r)$. Following corollary 2.31, for each $T \in \mathfrak{J}_r$ there exists a factor $*$-representation π_T of $C^*(A(r))$ such that $\pi_T(A(r)) = T$ and $S \approx T$ if and only if π_S and π_T are quasi-equivalent $*$-representations. Following Brian Davies [31] let $\mathfrak{A}(r)$ denote the smallest $*$-subalgebra of the enveloping von Neumann algebra of $C^*(A(r))$, which is closed under weak <u>sequential</u> convergence. Then the $*$-representation π of $C^*(A(r))$ defined by

$$\pi = \sum_{T \in \mathfrak{F}_r} \oplus \pi_T$$

has a unique extension π_σ to a σ-representation of the Σ^*-algebra $\mathfrak{A}(r)$ (theorem 3.1 of [31]). Further π_σ is a faithful σ-representation of $\mathfrak{A}(r)$ onto $\Sigma^*(\pi) = \Sigma^*(A(r))$ since π contains the reduced atomic representation of $C^*(A(r))$ (i.e., the direct sum of the irreducible representations of $C^*(A(r))$, taking one from each unitary equivalence class) (cf. theorem 3.2, page 152 of [31]). Thus in particular π_σ maps the center of $\mathfrak{A}(r)$ faithfully onto the center of $\Sigma^*(A(r))$. However Davies has defined his Borel structure on the quasi-spectrum $C^*(A(r))^\sim$ of $C^*(A(r))$ in such a way that the center of $\mathfrak{A}(r)$, and hence the center of $\Sigma^*(A(r))$ is isomorphic to the $*$-algebra of all bounded complex valued Davies-Borel functions on $C^*(A(r))^\sim$ (cf. theorem 4.2 of [32]). We describe this isomorphism more explicitly. If S is contained in the center of $\Sigma^*(A(r))$, then S acts on the space $\sum_{T \in \mathfrak{F}_r} \oplus \mathcal{H}(T)$ and for any particular T in $\mathfrak{F}_r$, the restriction of S to $\mathcal{H}(T)$ is an element of the center of $\mathcal{A}(T)$, the von Neumann algebra generated by T (cf. the last paragraph of the proof of theorem 3.4 of [33]). However since each T in $\mathfrak{F}_r$ is a factor operator, the restriction of S to $\mathcal{H}(T)$ is a complex multiple $f_S(T)$ of the identity operator on $\mathcal{H}(T)$. Further if T and T' are two quasi-equivalent elements of $\mathfrak{F}_r$, then there is a $*$-isomorphism φ of $\mathcal{A}(T)$ onto $\mathcal{A}(T')$ such that $\varphi(T) = T'$ (cf. proposition 1.34). Hence $f_S(T) = f_S(T')$ and we may therefore consider f_S to be a complex valued function defined on $\tilde{\mathfrak{F}}_r$. Then the set

$$\left\{f_S : S \in \text{center of } \Sigma^*(A(r))\right\}$$

is precisely the set of Davies-Borel functions on $\tilde{\mathfrak{F}}_r$ and the map $S \to f_S$ is an isomorphism of the center of $\Sigma^*(A(r))$ onto the $*$-algebra of Davies-Borel functions on $\widetilde{\mathfrak{F}}_r$ (cf. theorem 4.2 and its proof, of [32]). Hence the Davies-Borel sets are precisely those subsets B of $\widetilde{\mathfrak{F}}_r$ for which the operator

$$\sum_{T \in \mathfrak{F}_r} \oplus\ \chi_B(T^q)I_T$$

is contained in $\Sigma^*(A(r))$ (and hence in the center of $\Sigma^*(A(r))$).

The fact that the Mackey-Borel structure contains the Davies-Borel structure follows from theorem 4.1 of [31].

These general notions lead us to still one more characterization of smooth operators.

PROPOSITION 3.17. Let T be an operator. Then the following three conditions are equivalent.

1. T is a smooth operator.

2. The Davies-Borel structure and the Mackey-Borel structure coincide on $\hat{T}$.

3. The Davies-Borel structure and the Mackey-Borel structure coincide on $\widetilde{T}$.

PROOF. If T is smooth we have $\hat{T} = \widetilde{T}$ (theorem 2.32) and hence theorem 4.1 of [31], together with proposition 3.16 shows that 1 implies both 2 and 3. On the other hand Edward Effros has shown that if T is not smooth (and hence $C^*(T)$ is not GCR) then the Davies-Borel structure is properly contained in the Mackey-Borel structure. Thus 2 implies 1. On the other hand since $\hat{T}$ is a Mackey-Borel subset of $\widetilde{T}$, clearly 3 implies 2.

Proposition 3.16 serves as a dictionary connecting the Davies-Effros theory for representations of C^*-algebras to operator theory. With this

connection we may easily adapt Effros' characterization of the canonical
measures on the quasi-spectrum of a separable C^*-algebra [48] to obtain the
following characterization of the central measure classes on $\tilde{\mathfrak{F}}$. The first
three conditions of this theorem simply say that μ is a standard measure
class with essentially bounded domain. The crucial condition is of course
the fourth one.

THEOREM 3.18. (Effros) A measure class μ on $\tilde{\mathfrak{F}}$ is
central if and only if there exists a Mackey-Borel sub-
set B of $\tilde{\mathfrak{F}}$ such that

1. $\mu(\tilde{\mathfrak{F}} - B) = 0$.

2. $B \subset \tilde{\mathfrak{F}}_r$ for some $r > 0$.

3. B is a standard Borel space.

4. The Mackey and Davies Borel structures are the same
   when restricted to the set B.

PROOF. If μ is a central measure class, then there exists an operator
T in $\mathcal{L}$ such that

$$T \simeq \int_{\tilde{T}}^{\oplus} T(\xi)\, d\mu(\xi)$$

is the central decomposition of T. Then for $r = \|T\|$ we have $\mu(\tilde{\mathfrak{F}} - \tilde{\mathfrak{F}}_r) = 0$
and further

$$\pi_T \simeq \int_{C^*(A(r))^\sim}^{\oplus} \pi_T(\xi)\, d\mu(\xi)$$

is the central decomposition of π_T, where μ is now considered to be defined on $C^*(A(r))^\sim$ under the usual identification of $C^*(A(r))^\sim$ with
$\widetilde{A(r)} = \tilde{\mathfrak{F}}_r$ (proposition 3.16 and corollary 2.31). Here π_T and each $\pi_T(\xi)$
are $*$-representations of the C^*-algebra $C^*(A(r))$ such that $\pi_T(A(r)) = T$

and $\pi_T(\xi)(A(r)) = T(\xi)$. By [48] there exists a Mackey-Borel set B', $B' \subset C^*(A(r))^\sim$ which by corollary 2.31 corresponds to a Mackey-Borel set B of $\tilde{T} \subset \tilde{\mathfrak{F}}_r$ such that B' (and hence B by proposition 3.16 and corollary 2.31) is a standard Borel space with respect to both the Mackey and the Davies Borel structure and thus in particular, they must be the same Borel structure on B.

Conversely if a measure class μ on $\tilde{\mathfrak{F}}$ has the given properties, then μ restricted to $\tilde{\mathfrak{F}}_r$ is a central measure (when one identifies $\tilde{\mathfrak{F}}_r$ with $C^*(A(r))^\sim$ by proposition 3.16) and since $\mu(\tilde{\mathfrak{F}} - \tilde{\mathfrak{F}}_r) = 0$ we see that μ is a central measure class on $\tilde{\mathfrak{F}}$ (cf. theorem of [48]).

3. Multiplicity and the central decomposition

In this section we establish various technical properties of the central decomposition which we shall need in our spectral multiplicity theory to be developed in the next chapter.

LEMMA 3.19. Let Z be a Borel space and for each $m \in \Lambda$, a finite or countable set, let $\xi \to \mathcal{H}(\xi\,;m)$ denote a Borel field of Hilbert spaces (cf. definition 1, page 142 of [37]). Then

1) The map

$$\xi \to \sum_{m \in \Lambda} \oplus\, \mathcal{H}(\xi\,;m)$$

is a Borel field of Hilbert spaces on Z, and

2) If, for each m, $\xi \to S_m(\xi)$ is a Borel field of operators on Z such that $S_m(\xi) \in \mathcal{L}(\mathcal{H}(\xi\,;m))$, and if ν is a σ-finite standard Borel measure on Z and if there

is a positive number p such that $\|S_m(\xi)\| \leq p$ for all $m \in \Lambda$ and ν-almost all ξ, then

$$\xi \to \sum_{m \in \Lambda} \oplus S_m(\xi)$$

is a ν-essentially bounded Borel field of operators on Z (cf. definition 1, page 156 of [37]).

PROOF. In this section we shall base our methods on the development and terminology of Jacques Dixmier [37], preferring however to deal with Borel fields rather than measurable fields (cf. remark 3, page 143 of [37]).

Following definition 1, page 142 of [37] for each m there exists a fundamental sequence of Borel fields $\xi \to x_i^m(\xi)$, $i = 1,2,\ldots$. That is to say $\{x_i^m(\xi) : i = 1,2,\ldots\}$ is a sequence total in $\mathcal{H}(\xi\,;m)$ (i.e., $y \in \mathcal{H}(\xi\,;m)$ and $(y, x_i^m(\xi)) = 0$ for all i implies $y = 0$) and if $\xi \to y(\xi) \in \mathcal{H}(\xi\,;m)$ is a vector field on Z, then it is Borel if and only if $\xi \to (y(\xi), x_i^m(\xi))$ is Borel for all i (cf. proposition 2, page 144 of [37]). Further by the remark on page 142 of [37] we may assume $\|x_i^m(\xi)\| \leq 1$ for all i.

For each i, we define the vector field on Z as follows:

$$\xi \to x_i(\xi) = \sum_m \oplus 2^{-m/2} x_i^m(\xi) \in \sum_m \oplus \mathcal{H}(\xi\,;m).$$

Then

$$\begin{aligned}\xi \to &\left(x_i(\xi), x_j(\xi)\right) \\ &= \left(\sum_m \oplus 2^{-m/2} x_i^m(\xi), \sum_m \oplus 2^{-m/2} x_j^m(\xi)\right) \\ &= \sum_m 2^{-m}\left(x_i^m(\xi), x_j^m(\xi)\right)\end{aligned}$$

which is Borel since each $\xi \to \left(x_i^m(\xi), x_j^m(\xi)\right)$ is Borel (cf. remark following definition 1, page 142, of [37]) and a finite sum of Borel functions is Borel and the point-wise limit of a sequence of Borel functions is Borel (cf. bottom of page 90 of [115]). Further an easy verification shows that $\{x_i(\xi) : i = 1, 2, \ldots\}$ is total in $\mathcal{H}(\xi) = \sum_m \oplus \mathcal{H}(\xi\,;m)$. Thus by proposition 4, page 145 of [37] the field $\xi \to \mathcal{H}(\xi)$ is a Borel field of Hilbert spaces on Z and $\{\xi \to x_i(\xi) : i = 1, 2, \ldots\}$ is a fundamental vector field on Z.

Since $\|S_m(\xi)\| \leq p$ for all m and ν-almost all ξ we may form

$$\xi \to S(\xi) = \sum_m \oplus S_m(\xi)$$

and this field of operators on Z is ν-essentially bounded. Indeed $\|S(\xi)\| \leq p$ for ν-almost all ξ. To verify that $\xi \to S(\xi)$ is a Borel field we may apply proposition 1, page 156 of [37] which asserts that it is sufficient to verify that

$$\xi \to \big(S(\xi)\, x_i(\xi),\, x_j(\xi)\big)$$

is Borel for all i, j, where $\xi \to x_i(\xi)$ is a fundamental Borel field of vectors on Z. However by this same proposition we know that

$$\xi \to \big(S_m(\xi)\, x_i^m(\xi),\, x_j^m(\xi)\big)$$

is Borel for all i, j and m. Hence

$$\begin{aligned}\xi &\to \big(S(\xi)\, x_i(\xi),\, x_j(\xi)\big)\\ &\to \Big(\Big(\sum_m \oplus S_m(\xi)\Big)\Big(\sum_m \oplus 2^{-m/2} x_i^m(\xi)\Big),\, \sum_m \oplus 2^{-m/2} x_j^m(\xi)\Big)\\ &\to \sum_m 2^{-m}\big(S_m(\xi)\, x_i^m(\xi),\, x_j^m(\xi)\big)\end{aligned}$$

is Borel.

LEMMA 3.20. Suppose S and T are quasi-equivalent operators with central decompositions

$$S = \int_{\tilde{\mathfrak{F}}}^{\oplus} S(\xi)\, d\mu(\xi) \quad \text{and} \quad T = \int_{\tilde{\mathfrak{F}}}^{\oplus} T(\xi)\, d\mu(\xi)$$

then

$$S \oplus T \simeq \int_{\tilde{\mathfrak{F}}}^{\oplus} \Big(S(\xi) \oplus T(\xi)\Big)\, d\mu(\xi)$$

is the central decomposition of $S \oplus T$.

PROOF. Of course the relative complement of $\tilde{S} = \tilde{T}$ in $\tilde{\mathfrak{F}}$ is a μ-null set. The unitary equivalence is effected by the obvious canonical isomorphism between the Hilbert space

$$\int_{\tilde{\mathfrak{F}}}^{\oplus} \mathcal{H}(S(\xi))\, d\mu(\xi) \;\oplus\; \int_{\tilde{\mathfrak{F}}}^{\oplus} \mathcal{H}(T(\xi))\, d\mu(\xi)$$

and the Hilbert space

$$\int_{\tilde{\mathfrak{F}}}^{\oplus} \Big[\mathcal{H}(S(\xi)) \oplus \mathcal{H}(T(\xi))\Big]\, d\mu(\xi)$$

and we shall find it convenient during the proof to simply identify these two spaces.

By the previous lemma 3.19, we have that $\xi \to S(\xi) \oplus T(\xi)$ is a μ-essentially bounded Borel field of operators such that $S(\xi) \oplus T(\xi)$ is contained in the quasi-equivalence class ξ and

$$S \oplus T \simeq \int_{\tilde{\mathfrak{F}}}^{\oplus} \Big(S(\xi) \oplus T(\xi)\Big)\, d\mu(\xi).$$

We must verify that this is indeed the central decomposition, i.e., that the algebra of diagonalizable operators is precisely the center of $\mathcal{A}(S \oplus T)$. Since any von Neumann algebra is generated by its projections, it is sufficient to verify that the projections associated with characteristic functions of Borel subsets of $\tilde{\mathfrak{F}}$ are precisely the projections in the center of $\mathcal{A}(S \oplus T)$. Let B denote a Borel subset of $\tilde{\mathfrak{F}}$. By the correspondence between quasi-equivalence classes of operators and central measures on $\tilde{\mathfrak{F}}$ (theorem 3.8) we have that

$$\int_B^{\oplus} S(\xi)\, d\mu(\xi) \approx \int_B^{\oplus} T(\xi)\, d\mu(\xi)$$

and

$$\int_{\tilde{\mathfrak{F}}-B}^{\oplus} S(\xi)\, d\mu(\xi) \approx \int_{\tilde{\mathfrak{F}}-B}^{\oplus} T(\xi) d\mu(\xi).$$

Further since the given decompositions are central we have (proposition 1.35, part 3)

$$\int_B^{\oplus} S(\xi)\, d\mu(\xi) \;\underset{\circ}{|}\; \int_{\tilde{\mathfrak{F}}-B}^{\oplus} S(\xi)\, d\mu(\xi)$$

and

$$\int_B^{\oplus} T(\xi)\, d\mu(\xi) \;\underset{\circ}{|}\; \int_{\tilde{\mathfrak{F}}-B}^{\oplus} T(\xi)\, d\mu(\xi).$$

Since

$$\int_B^{\oplus} \Big(S(\xi) \oplus T(\xi)\Big) d\mu(\xi) \simeq \int_B^{\oplus} S(\xi)\, d\mu(\xi) \oplus \int_B^{\oplus} T(\xi)\, d\mu(\xi)$$

(and a similar statement with B replaced by $\tilde{\mathfrak{F}} - B$) we conclude that

$$\int_B^{\oplus} \Big(S(\xi) \oplus T(\xi)\Big) d\mu(\xi) \;\underset{\circ}{|}\; \int_{\tilde{\mathfrak{F}}-B}^{\oplus} \Big(S(\xi) \oplus T(\xi)\Big) d\mu(\xi)$$

for all Borel sets $B \subset \tilde{\mathfrak{F}}$. We thus have that all the diagonalizable projections, and hence all the diagonalizable operators are contained in the center of $\mathcal{A}(S \oplus T)$ (cf. proposition 1.35, part 3).

Thus

$$\int_{\tilde{\mathfrak{F}}}^{\oplus} \mathcal{A}\big(S(\xi) \oplus T(\xi)\big)\, d\mu(\xi)$$

is a von Neumann algebra generated by the operator $S \oplus T$ corresponding to the Borel field $\xi \to S(\xi) \oplus T(\xi)$ and the diagonalizable operators. But since the diagonalizable operators are contained in the center of $\mathcal{A}(S \oplus T)$ we have

$$\mathcal{A}(S \oplus T) = \int_{\tilde{\mathfrak{F}}}^{\oplus} \mathcal{A}\big(S(\xi) \oplus T(\xi)\big)\, d\mu(\xi).$$

Now if E is a projection in the center of $\mathcal{A}(S \oplus T)$, it is of the form

$$E = \int_{\tilde{\mathfrak{F}}}^{\oplus} E(\xi)\, d\mu(\xi)$$

where $E(\xi)$ is a projection in the center of $\mathcal{A}\big(S(\xi) \oplus T(\xi)\big)$ for μ - almost all ξ (cf. theorem 4, page 176 and theorem 3, page 174 of [37]). However $S(\xi)$ and $T(\xi)$ are quasi-equivalent factor operators and hence $S(\xi) \oplus T(\xi)$ is also a factor operator (contained in the class ξ) and thus $\mathcal{A}\big(S(\xi) \oplus T(\xi)\big)$ is a factor von Neumann algebra. Hence for μ - almost all ξ in $\tilde{\mathfrak{F}}$, either $E(\xi)$ is the zero operator, or the identity operator $I(\xi)$ on $\mathcal{H}(S(\xi)) \oplus \mathcal{H}(T(\xi))$. Let $B = \{\xi : E(\xi) = I(\xi)\}$ which is a Borel subset of $\tilde{\mathfrak{F}}$. Thus the projection E is a diagonalizable operator since it corresponds to the characteristic function of the set B. Thus the center of $\mathcal{A}\big(S(\xi) \oplus T(\xi)\big)$ is precisely the set of diagonalizable operators and hence

$$S \oplus T = \int_{\tilde{\mathfrak{F}}}^{\oplus} \big[S(\xi) \oplus T(\xi)\big]\, d\mu(\xi)$$

is the central decomposition of $S \oplus T$.

Lemma 3.20 has a obvious generalized form (cf. lemma 3.19) which we shall use from time to time.

LEMMA 3.21. Let $\{S_m\}$ be a countable (finite or infinite) collection of quasi-equivalent operators with central decompositions

$$S_m = \int_{\tilde{\mathfrak{F}}}^{\oplus} S_m(\xi)\, d\mu(\xi).$$

Then

$$\sum_m \oplus S_m \simeq \int_{\tilde{\mathfrak{F}}}^{\oplus} \Big(\sum_m \oplus S_m(\xi) \Big) d\mu(\xi)$$

is the central decomposition of $\sum_m \oplus S_m$.

COROLLARY 3.22. If

$$S \simeq \int_{\tilde{S}}^{\oplus} S(\xi)\, d\mu(\xi)$$

is the central decomposition of S and n is a positive integer or $\aleph_0$, then $\xi \to \textcircled{n} S(\xi)$ is a Borel field of operators and

$$\textcircled{n} S \simeq \int_{\tilde{S}}^{\oplus} \textcircled{n} S(\xi)\, d\mu(\xi)$$

is the central decomposition of $\textcircled{n} S$.

PROPOSITION 3.23. Let S and T denote two operators with central decompositions

$$S \simeq \int_{\tilde{\mathfrak{F}}}^{\oplus} S(\xi)\, d\mu(\xi) \quad \text{and} \quad T \simeq \int_{\tilde{\mathfrak{F}}}^{\oplus} T(\xi)\, d\nu(\xi).$$

Then $S \lesssim T$ if and only if $\mu << \nu$ and $S(\xi) \lesssim T(\xi)$ for μ - almost all ξ in $\tilde{\mathfrak{F}}$.

PROOF. Suppose first that $S \lesssim T$, i.e., that S is unitarily equivalent to a suboperator of T. Then in particular $T \} S$ and hence by proposition 3.10 we have $\mu << \nu$. Further we may express $\nu = \mu \vee \mu'$ where $\mu \perp \mu'$ (cf. theorem 2, page 80 of [77]). Let

$$T_1 = \int_{\tilde{\mathfrak{F}}}^{\oplus} T(\xi)\, d\mu(\xi) \quad \text{and} \quad T_2 = \int_{\tilde{\mathfrak{F}}}^{\oplus} T(\xi)\, d\mu'(\xi).$$

Then $T \simeq T_1 \oplus T_2$ and $T_1 \overset{|}{\circ} T_2$ and $S \overset{|}{\circ} T_2$. Since $S \lesssim T$ we have $S \lesssim T_1$. Thus there exists a projection E in $\mathcal{A}(T_1)'$ such that $S \simeq T_1 E$. By part 4 of theorem 3.4, E has a decomposition

$$E = \int_{\tilde{\mathfrak{F}}}^{\oplus} E(\xi)\, d\mu(\xi)$$

where $E(\xi)$ is a projection in $\mathcal{A}(T(\xi))'$ for μ - almost all ξ. By proposition 3, page 159 of [37] we have $S(\xi) \simeq T(\xi)\, E(\xi)$ for μ - almost all ξ, i.e., $S(\xi) \lesssim T(\xi)$ for μ - almost all ξ.

Conversely suppose $\mu << \nu$ and $S(\xi) \lesssim T(\xi)$ for μ - almost all ξ. Again we may write $\nu = \mu \vee \mu'$ where $\mu \perp \mu'$. Thus it will be sufficient to prove

$$\int_{\tilde{\mathfrak{F}}}^{\oplus} S(\xi)\, d\mu(\xi) \lesssim \int_{\tilde{\mathfrak{F}}}^{\oplus} T(\xi)\, d\mu(\xi).$$

For this purpose we consider the Borel field of operators $\xi \to S(\xi) \oplus T(\xi)$ (cf. lemma 3.19). Since $S(\xi)$ and $T(\xi)$ are both in the same quasi-equivalence class ξ we have that $S(\xi) \oplus T(\xi)$ is also in this quasi-equivalence class ξ. Hence if we let

$$T_1 = \int_{\tilde{\mathfrak{F}}}^{\oplus} T(\xi)\, d\mu(\xi)$$

then the integral is the central decomposition of T_1, $T_1 \lesssim T$, T_1 is quasi-equivalent to S and by lemma 3.20, the central decomposition of

$S \oplus T_1$ is given by

$$S \oplus T_1 \simeq \int_{\tilde{\mathfrak{F}}}^{\oplus} \big(S(\xi) \oplus T(\xi)\big)\, d\mu(\xi)$$

and by theorem 3.4, part 3, we have

$$\mathcal{A}(S \oplus T_1) \simeq \int_{\tilde{\mathfrak{F}}}^{\oplus} \mathcal{A}\big(S(\xi) \oplus T(\xi)\big)\, d\mu(\xi).$$

By proposition 1.35, part 2, the assertion that $S(\xi) \precsim T(\xi)$ is equivalent to the assertion that $E(\xi) \precsim F(\xi)$ relative to $\mathcal{A}\big(S(\xi) \oplus T(\xi)\big)'$, where $E(\xi)$ is the projection of $\mathcal{H}\big(S(\xi) \oplus T(\xi)\big)$ onto $\mathcal{H}(S(\xi))$ and $F(\xi)$ is the projection of $\mathcal{H}\big(S(\xi) \oplus T(\xi)\big)$ onto $\mathcal{H}(T(\xi))$. Further $E = \int_{\tilde{\mathfrak{F}}}^{\oplus} E(\xi)\, d\mu(\xi)$ corresponds to the projection of $\mathcal{H}(S) \oplus \mathcal{H}(T_1)$ onto $\mathcal{H}(S)$. Similarly $F = \int_{\tilde{\mathfrak{F}}}^{\oplus} F(\xi)\, d\mu(\xi)$ corresponds to the projection of $\mathcal{H}(S) \oplus \mathcal{H}(T_1)$ onto $\mathcal{H}(T_1)$. By problem 15, page 229 of [37] we have that $E(\xi) \precsim F(\xi)$ (re $\mathcal{A}(S(\xi) \oplus T(\xi))'$) for μ-almost all ξ implies that $E \precsim F$ (re $\mathcal{A}(S \oplus T_1)'$) and hence again by part 2 of proposition 1.35 that $S \precsim T_1 \precsim T$.

COROLLARY 3.24. If S and T are quasi-equivalent operators with central decompositions

$$S \simeq \int_{\tilde{\mathfrak{F}}}^{\oplus} S(\xi)\, d\mu(\xi) \qquad \text{and} \qquad T \simeq \int_{\tilde{\mathfrak{F}}}^{\oplus} T(\xi)\, d\mu(\xi)$$

then $S \simeq T$ is and only if $S(\xi) \simeq T(\xi)$ for μ-almost all ξ.

COROLLARY 3.25. If T is an operator of continuous type, with central decomposition

$$T \simeq \int_{\tilde{T}}^{\oplus} T(\xi)\, d\mu(\xi)$$

and λ is any positive real number or $\aleph_0$, then the central decomposition of $(\lambda) T$ has the form

$$(\lambda) T \simeq \int_{\tilde{T}}^{\oplus} (\lambda) T(\xi)\, d\mu(\xi).$$

PROOF. By part 6 of theorem 3.4, μ - almost all the component factor operators $T(\xi)$ are of continuous type. Suppose $T \simeq (2) S$ for an operator S (cf. proposition 1.24). Then $T \approx S$ and S has a central decomposition of the form

$$S \simeq \int_{\tilde{T}}^{\oplus} S(\xi)\, d\mu(\xi)$$

by theorem 3.8. Hence by corollary 3.22 we have

$$T \simeq (2) S \simeq \int_{\tilde{T}}^{\oplus} (2) S(\xi)\, d\mu(\xi)$$

where this is the central decomposition of $(2) S$. By corollary 3.24 we have that

$$(2) S(\xi) \simeq T(\xi)$$

for μ - almost all ξ. Thus $S \simeq (\tfrac{1}{2}) T$ if and only if $S(\xi) \simeq (\tfrac{1}{2}) T(\xi)$ for μ - almost all ξ. Thus

$$(\tfrac{1}{2}) T \simeq \int_{\tilde{T}}^{\oplus} (\tfrac{1}{2}) T(\xi)\, d\nu(\xi)$$

is the central decomposition of $(\tfrac{1}{2}) T$. This result, applied repeatedly together with corollary 3.22, establishes our result in the special case where λ is a dyadic rational, i.e., a positive number of the form $m2^{-n}$ where m and n are integers. (Note that the case λ is $\aleph_0$ is already established for arbitrary operators in corollary 3.22.) Next let λ be any positive real number and let $\lambda = \sum_{i=1}^{\infty} r_i$ be a series representation of λ in terms of dyadic rationals r_i. By our result for dyadic rationals and lemma 3.21, we have that the central decomposition of $\textcircled{\lambda}T$ (cf. proposition 1.41) is given by

$$\begin{aligned}
\textcircled{\lambda}T \simeq \sum_{i=1}^{\infty} \oplus \textcircled{r_i} T &\simeq \sum_{i=1}^{\infty} \oplus \textcircled{r_i} \int_{\tilde{T}}^{\oplus} T(\xi)\, d\mu(\xi) \\
&\simeq \sum_{i=1}^{\infty} \oplus \int_{\tilde{T}}^{\oplus} \textcircled{r_i} T(\xi)\, d\mu(\xi) \\
&\simeq \int_{\tilde{T}}^{\oplus} \sum_{i=1}^{\infty} \oplus \textcircled{r_i} T(\xi)\, d\mu(\xi) \\
&\simeq \int_{\tilde{T}}^{\oplus} \textcircled{\lambda}T(\xi)\, d\mu(\xi).
\end{aligned}$$

Chapter 4

SPECTRAL MULTIPLICITY THEORY AND THE UNITARY EQUIVALENCE PROBLEM FOR OPERATORS

One may use proposition 2.35 to determine operators up to weak equivalence and theorem 3.8 to distinguish operators, up to quasi-equivalence, within a weak equivalence class. We now turn to the problem of distinguishing operators up to unitary equivalence, within a quasi-equivalence class. We shall do this by imitating and generalizing the spectral multiplicity theory for normal operators as described in Paul Halmos' classic text [77]. Thus if a quasi-equivalence class $t \in \tilde{\mathcal{L}}$ corresponds to a central measure class μ in $C\mathcal{M}(\tilde{\mathfrak{Z}})$, then the operators within the quasi-equivalence class t will be distinguished, up to unitary equivalence, by spectral multiplicity functions whose domain is the set of finite standard measure classes on $\tilde{\mathfrak{Z}}$ which are absolutely continuous with respect to μ. Our definition of "multiplicity function" will differ in two respects, one minor and one major, from the multiplicity functions considered by Paul Halmos and defined on pages 80 and 81 of [77]. The minor difference is that we shall define the value of the multiplicity function, at the identically zero measure, to be $+\infty$. (Halmos defined this value to be 0.) With this change our multiplicity functions will be monotonically nonincreasing on the ideal of measure classes absolutely continuous with respect to μ. The major difference is that for Halmos the values of a multiplicity function are cardinal numbers (finite or infinite) whereas in our theory the values will be nonnegative real numbers and $\aleph_0$ (which we shall denote simply $+\infty$). Presumably a theory developed for operators on a nonseparable space would require distinguishing different infinite cardinals. This is unnecessary in our case as the entire theory of this memoir is being developed only for operators on separable Hilbert spaces. We proceed with the formal definition.

DEFINITION 4.1. If μ is a central measure class on $\tilde{\mathfrak{F}}$, we define a <u>multiplicity function for</u> μ to be a function f defined on the lattice ideal of all finite measure classes ν on $\tilde{\mathfrak{F}}$ such that $\nu << \mu$, having nonnegative real numbers and $+\infty$ for possible values, and satisfying the following four axioms:

1. If $\nu \equiv 0$, then $f(\nu) = +\infty$.

2. If ν_1, ν_2 are finite measure classes on $\tilde{\mathfrak{F}}$ and $\nu_1 << \nu_2 << \mu$, then $f(\nu_2) \leq f(\nu_1)$.

3. If $\nu << \mu$ and ν is a finite measure class on $\tilde{\mathfrak{F}}$ which is the supremum of a countable family $\{\nu_j\}$ of two-by-two orthogonal standard measure classes on $\tilde{\mathfrak{F}}$, then

$$f(\nu) = \inf\left\{f(\nu_j)\right\}.$$

4. If $f(\nu) = 0$, then there exists a nonzero finite measure class ν' such that $\nu' << \nu$ and $f(\nu') > 0$.

REMARK 4.2. We remark that condition 3 of the previous definition (in the presence of the other axioms) is equivalent to the same condition with the adjective "two-by-two orthogonal" removed. Indeed assume conditions 2 and 3 above and let $\{\nu_j\}$ denote any countable collection of finite measure classes on $\tilde{\mathfrak{F}}$ such that $\nu_j << \mu$ and $\nu = \sup\{\nu_j\}$. Then, by theorem 2, page 80 of [77] we can find a sequence of measure classes $\{\nu_j'\}$ such that $\nu_j' << \nu_j$ for each j, $\sup\{\nu_j\} = \sup\{\nu_j'\}$ and the ν_j' are two-by-two orthogonal. Since $\nu_j' << \nu_j << \nu$ we have $f(\nu_j') \geq f(\nu_j) \geq f(\nu)$ for all j and hence

$$f(\nu) = \inf_j\left\{f(\nu_j')\right\} \geq \inf_j\left\{f(\nu_j)\right\} \geq f(\nu)$$

and thus

$$f(\nu) = \inf_j \left\{ f(\nu_j) \right\} .$$

The last condition (4) is a technical one which is needed to ensure that f is basically positive in the sense that

$$\mu = \vee \left\{ \nu : f(\nu) > 0 \right\} .$$

We of course intend to associate such a multiplicity function with each unitary equivalence class within a quasi-equivalence class, and for that purpose we must first prove a couple of technical lemmas which are long on hypotheses and short on conclusions.

LEMMA 4.3 (Cf. theorem 2.8 of [94]) Let T be a type I operator with central decomposition

$$T \simeq \int_{\tilde{\mathfrak{F}}}^{\oplus} T(\xi)\, d\mu(\xi) .$$

By definitions 1.21 and 1.28, T is quasi-equivalent to a multiplicity free operator S and by theorem 3.8 S has a central decomposition of the form

$$S \simeq \int_{\tilde{\mathfrak{F}}}^{\oplus} S(\xi)\, d\mu(\xi) .$$

By part 5 of theorem 3.4, μ - almost all of the component operators $S(\xi)$ are irreducible. Thus there exists a Borel subset $B \subset \tilde{\mathfrak{F}}$ whose relative complement in $\tilde{\mathfrak{F}}$ is a μ - null set and such that $S(\xi)$ is an irreducible operator for each ξ in B. Since $T(\xi)$ and $S(\xi)$ are quasi-equivalent type I factor operators (they are both in the same class ξ) the multiplicity theory for type I factor operators (theorem 1.43) implies that for each $\xi \in B$, there exists $n(\xi)$, a positive integer or $+\infty$, such that $T(\xi) \simeq n(\xi)\, S(\xi)$.

Conclusion: The map $\xi \to n(\xi)$ is a Borel map on B.

PROOF. We have $\xi \to T(\xi)$ and $\xi \to S(\xi)$ are Borel maps on B. Hence by proposition 2.10 the map

$$\xi \to \mathcal{R}\big(T(\xi), S(\xi)\big) \to \dim \mathcal{R}\big(T(\xi), S(\xi)\big) = n(\xi)$$

is a Borel map on B.

LEMMA 4.4 (Cf. lemma 18, page 464 of [127]) Let T be a type II operator with central decomposition

$$T \simeq \int_{\tilde{\mathfrak{F}}}^{\oplus} T(\xi)\, d\mu(\xi).$$

Then by definition of type II (definition 1.28) T is semi-finite and hence quasi-equivalent to some finite type II operator S. Since T and S are quasi-equivalent theorem 3.8 implies the central decomposition of S has the form

$$S \simeq \int_{\tilde{\mathfrak{F}}}^{\oplus} S(\xi)\, d\mu(\xi).$$

Thus by part 6 of theorem 3.4, there exists a Borel subset B of $\tilde{\mathfrak{F}}$, whose relative complement in $\tilde{\mathfrak{F}}$ is a μ-null set, and such that $S(\xi)$ is a finite type II factor operator for every ξ in B. By the multiplicity theory for type II factor operators (theorem 1.43) and the fact that $T(\xi) \approx S(\xi)$ for all ξ in B, there exists $m(\xi)$, a positive real number or $+\infty$, such that

$$T(\xi) \simeq \textcircled{m(\xi)}\, S(\xi)$$

for every ξ in B.

Conclusion: The map $\xi \to m(\xi)$ of B into $R^+ \cup \{+\infty\}$ is a Borel map.

PROOF. By corollary 3.25, for each $\lambda > 0$, the central decomposition of $\textcircled{\lambda} S$ is given by

$$\textcircled{\lambda} S \simeq \int_{\tilde{\mathfrak{Z}}}^{\oplus} \textcircled{\lambda} S(\xi)\, d\mu(\xi)\,.$$

By proposition 1.12 and part 3 of proposition 1.35 there exist central projections E and F in $\mathcal{A}(\textcircled{\lambda} S)$ and $\mathcal{A}(T)$ respectively such that $TF \lesssim \textcircled{\lambda} SE$ and $\textcircled{\lambda} S(I-E) \lesssim T(I-F)$. Further since $T \approx \textcircled{\lambda} S$ we have that both central projections are diagonalizable operators which correspond to the characteristic function of the <u>same</u> Borel subset $B \subset \tilde{T} = \tilde{S} \subset \tilde{\mathfrak{Z}}$ (since $TF \approx \textcircled{\lambda} SE$ and $T(I-F) \approx \textcircled{\lambda} S(I-E)$). Hence by proposition 3.23 we have

$$T(\xi) \simeq \textcircled{m(\xi)}\, S(\xi) \lesssim \textcircled{\lambda} S(\xi)$$

for μ - almost all ξ in B, and

$$T(\xi) \simeq \textcircled{m(\xi)}\, S(\xi) \gtrsim \textcircled{\lambda} S(\xi)$$

for μ - amost all ξ in the relative complement of B.

Let $\{\lambda_i\}$ denote an increasing sequence of positive real numbers, $\lambda_i \neq \lambda$, approaching λ from the left. Then for each i, there exists a Borel set B_i such that $m(\xi) \leq \lambda_i$ for all ξ in B_i and $m(\xi) \geq \lambda_i$ for all ξ in $\tilde{\mathfrak{Z}} - B_i$. Let $B = \bigcup_i B_i$. Then B is a Borel set and $m(\xi) < \lambda$

if and only if $\xi \in B$. Thus the map $\xi \to m(\xi)$ is Borel.

DEFINITION 4.5. Let t be a fixed element of $\tilde{\mathcal{L}}$, i.e., a quasi-equivalence class of operators acting on a separable Hilbert space and let μ denote the central measure class on $\tilde{\mathfrak{F}}$ corresponding to t (cf. theorem 3.8 and proposition 3.10). To each (unitary equivalence class of an) operator T in the quasi-equivalence class t we shall associate a multiplicity function f_T defined on the finite measure classes on $\tilde{\mathfrak{F}}$ absolutely continuous with respect to μ, by a procedure which we now outline.

Recall that we have introduced the notation δ_o for an added distinguished point of $\tilde{\mathcal{L}}$ which has the property that $s \,\}\, \delta_o$ for every $s \in \tilde{\mathcal{L}}$ (cf. remarks preceding proposition 1.57). Further by proposition 1.34 and corollary 1.39 we may translate proposition 1.29 into a statement about quasi-equivalence classes. Thus t may be expressed uniquely in the form

$$t = t_I \vee t_{II} \vee t_{III}$$

such that if $i \neq j$ and $t_i \neq \delta_o$ and $t_j \neq \delta_o$, then $t_i \overset{|}{\circ} t_j$, $i,j = I, II, III$, and such that either $t_i = \delta_o$ or t_i is a quasi-equivalence class of operators of type i, $i = I$, II or III.

Since the correspondence between $\tilde{\mathcal{L}}$ and $C\mathcal{M}(\tilde{\mathfrak{F}})$ is a lattice isomorphism (proposition 3.10) it follows that the measure class μ also may be expressed

$$\mu = \mu_I \vee \mu_{II} \vee \mu_{III}$$

where μ_i is the central measure class corresponding to t_i, $i = I$, II or III, and $\mu_i \perp \mu_j$ if $i \neq j$, $i,j = I, II, III$. We shall call μ_i the type i part of μ for $i = I, II, III$. Then the central decomposition of T may be written

$$T = \sum_{i=I}^{III} \oplus \int_{\tilde{\mathfrak{F}}}^{\oplus} T(\xi)\, d\mu_i(\xi) .$$

Similarly, if $\nu \in C\mathcal{M}(\tilde{\mathfrak{F}})$ and $\nu << \mu$ then we may decompose ν

$$\nu = \nu_I \vee \nu_{II} \vee \nu_{III}$$

where $\nu_i << \mu_i$ for $i = I, II, III$.

By property 3 of definition 4.1 we know that any multiplicity function f for μ must satisfy the condition

$$f(\nu) = \min\left\{ f(\nu_i) : i = I,\ II,\ III \right\}.$$

Hence we can (and will) define f_T separately for type I, type II and type III measures absolutely continuous with respect to μ, and then use this condition to extend it to all measure classes absolutely continuous with respect to μ. Of course if ν is the identically zero measure class it corresponds to δ_o and condition 1 of definition 4.1 forces us to define $f_T(\nu) = +\infty$.

The problem of distinguishing representations up to unitary equivalence within a type III quasi-equivalence class is trivial (proposition 1.20). Thus we shall easily dispose of this case by defining $f_T(\nu_{III}) = +\infty$ for any type III central measure ν_{III} absolutely continuous with respect to μ. (Of course a central measure is called type i, $i = I$, II or III if it is the measure in the central decomposition of a type i operator, $i = I$, II or III.)

The multiplicity theory for type I measures is remarkably similar to the spectral multiplicity theory devised for normal operators [77]. Let T_I denote the type I part of T (proposition 1.29). Then its central decomposition has the form

$$T_I = \int_{\tilde{\mathfrak{F}}}^{\oplus} T(\xi)\, d\mu_I(\xi)$$

where $T(\xi)$ is a type I operator for μ_I - almost all ξ (theorem 3.4, part 6). By theorem 1.43 we have, for μ_I - almost all ξ, that there exists an irreducible operator $T_o(\xi)$ such that $T(\xi) \simeq n(\xi)\, T_o(\xi)$ where $n(\xi)$ is a positive integer or $+\infty$. Then for any $\nu_I << \mu_I$, $\nu_I \neq 0$, we define

$$\begin{aligned} f_T(\nu_I) &= \nu_I\text{- ess. inf.}\left\{n(\xi)\right\} \\ &= \operatorname{Sup}\left\{n(\xi') : \xi' \in \tilde{\mathfrak{F}} \text{ and } \nu_I\{\xi : n(\xi) < n(\xi')\} = 0\right\}. \end{aligned}$$

We next turn our attention to the definition of the multiplicity function f_T for type II measures. The type II operator T_{II} has central decomposition

$$T_{II} \simeq \int_{\tilde{\mathfrak{F}}}^{\oplus} T(\xi)\, d\mu_{II}(\xi) .$$

Since T is type II and hence semi-finite there exists a finite operator S which is quasi-equivalent to T_{II}, i.e., $S \in t_{II}$. Since $S \approx T_{II}$, theorem 3.8 implies that S has a central decomposition of the form

$$S \simeq \int_{\tilde{\mathfrak{F}}}^{\oplus} S(\xi)\, d\mu_{II}(\xi).$$

Thus $S(\xi)$ and $T(\xi)$ are quasi-equivalent operators (they are both in the same quasi-equivalence class ξ) and by part 6 of theorem 3.4, $S(\xi)$ is a finite operator for μ_{II} - almost all ξ. By the multiplicity theory for type II factor operators (theorem 1.43) $T(\xi) \simeq m(\xi)\, S(\xi)$ for μ_{II} - almost all ξ, where $m(\xi)$ is a positive real number or $+\infty$. Thus for $\nu_{II} << \mu_{II}$, $\nu_{II} \neq 0$, we define

$$\begin{aligned} f_T(\nu_{II}) &= \nu_{II}\text{- ess. inf.}\left\{m(\xi)\right\} \\ &= \operatorname{Sup}\left\{m(\xi') : \xi' \in \tilde{\mathfrak{F}} \text{ and } \nu_{II}\{\xi : m(\xi) < m(\xi')\} = 0\right\} . \end{aligned}$$

REMARK 4.6. We have finally finished defining the multiplicity function f_T associated with an operator T. Note that if the type II part of T is nontrivial (i.e., $t_{II} \neq \delta_o$) then our definition of the multiplicity function f_T depended on the choice of a finite operator in the type II part t_{II} of t. In this case we call f_T the relative multiplicity function for T and indicate its dependence on this choice of finite operator S quasi-equivalent to the type II part T_{II} of T by the notation $f_T = f_T^S$. We could get rid of this relativity in the type II case if only there were a canonical way of choosing a finite operator in the type II part t_{II} of t. Since T_{II} is semi-finite the quasi-equivalence class t_{II} contains precisely one (up to unitary equivalence) standard operator (proposition 1.33). Unfortunately this nice canonical choice need not be finite. However we can remove <u>some</u> of the relativity of the relative multiplicity function by agreeing that <u>if</u> the standard operator in the quasi-equivalence class t_{II} is finite, then the above defined multiplicity function is to be taken with respect to that standard operator. Even if the standard operator S in t_{II} is not finite, it does have a unique decomposition, $S \simeq S_1 \oplus S_2$ where $S_1 \overset{|}{\circ} S_2$, S_1 is finite and S_2 is infinite (proposition 1.16). In this case we can agree to define our relative multiplicity function with respect to a finite operator S' in t_{II} which is unitarily equivalent to $S_1 \oplus S_2'$ where $S_1 \overset{|}{\circ} S_2'$ and S_2' is a finite operator quasi-equivalent to the infinite semi-finite operator S_2. This convention will at least remove some of the relativity in the definition of the relative multiplicity function.

LEMMA 4.7. Let T be an operator with associated central measure class μ on $\tilde{\mathfrak{J}}$ and let S denote a finite operator quasi-equivalent to the type II part T_{II} of T. Then the relative multiplicity function f_T^S defined above, on the finite measure classes absolutely continuous with respect to μ, satisfies the axioms of definition 4.1.

PROOF. The first three conditions of definition 4.1 are obvious verifications and we prove only that f_T^S satisfies axiom 4. Suppose $f_T^S(\nu) = 0$. Then $\nu \not\equiv 0$ since by definition if $\nu \equiv 0$, then $f_T^S(\nu) = +\infty$. Write $\nu = \nu_I \vee \nu_{II} \vee \nu_{III}$. By definition $f_T^S(\nu_I) \geq 1$ and $f_T^S(\nu_{III}) = +\infty$. Thus $0 = f_T^S(\nu) = \inf\left\{f_T^S(\nu_i) : i = I, II, III\right\}$ implies that $f_T^S(\nu_{II}) = 0$. Hence

$$0 = f_T^S(\nu) = \nu_{II}\text{- ess. inf.} \left\{m(\xi)\right\} .$$

By the multiplicity theory for type II factor operators, $m(\xi) > 0$ for μ_{II}-almost all ξ. Let B_o denote a Borel subset of $\tilde{\mathfrak{F}}$ such that $m(\xi)$ is defined (and hence positive) for all ξ in B_o, and $\mu_{II} = \mu_{B_o}$ where μ_{B_o} is defined by $\mu_{B_o}(C) = \mu(B_o \cap C)$ for all Borel subsets C of $\tilde{\mathfrak{F}}$. Since $\nu_{II} << \mu_{II}$ there exists a Borel set B such that $B \subset B_o$ and $\nu_{II} \sim \mu_B$. Since $\xi \to m(\xi)$ is a Borel function on B_o (lemma 4.4) we have, for each positive rational number r, that

$$B_r = \left\{\xi : \xi \in B_o \text{ and } m(\xi) > r\right\}$$

is a Borel set and $B_o = \cup_r B_r$. If $\mu(B \cap B_r) = 0$ for every positive rational r, then $\mu(B) = \mu\left(\cup_r (B \cap B_r)\right) = 0$ which would imply that $\nu_{II} \sim \mu_B = 0$ which is impossible since $f_T^S(\nu_{II}) = 0$ and $f_T^S(0) = +\infty$. Thus there exists a positive rational number r such that $\mu(B \cap B_r) \neq 0$ and hence $\mu_{B \cap B_r} \neq 0$. Thus $\mu_{B \cap B_r}$ is a non-zero finite measure such that $\mu_{B \cap B_r} << \mu_B \sim \nu_{II} << \nu$ and $f_T^S(\mu_{B \cap B_r}) \geq r > 0$.

THEOREM 4.8. Let S and T denote quasi-equivalent operators and let μ denote the central measure class on $\tilde{\mathfrak{F}}$ corresponding to this quasi-equivalence class. Let R denote any finite type II operator quasi-equivalent to the type II part of S. Let f_S^R and f_T^R denote the relative multiplicity functions for S and T respectively. Then $S \lesssim T$ if and only if $f_S^R(\nu) \leq f_T^R(\nu)$ for all finite measure classes ν on $\tilde{\mathfrak{F}}$, $\nu << \mu$.

PROOF. Let $\nu << \mu$ and let $\mu = \mu_I \vee \mu_{II} \vee \mu_{III}$ and $\nu = \nu_I \vee \nu_{II} \vee \nu_{III}$ denote the decomposition of ν and μ into their type I, II and III parts respectively (cf. propositions 1.29, 1.34, corollary 1.39 and proposition 3.10 in that order). Let

$$S \simeq \int_{\tilde{\mathfrak{F}}}^{\oplus} S(\xi)\, d\mu(\xi), \qquad T \simeq \int_{\tilde{\mathfrak{F}}}^{\oplus} T(\xi)\, d\mu(\xi) \qquad \text{and} \qquad R \simeq \int_{\tilde{\mathfrak{F}}}^{\oplus} R(\xi)\, d\mu_{II}(\xi)$$

denote the central decompositions of S, T and R respectively. For μ_I-almost all ξ, there exists an irreducible operator $N(\xi)$ such that $S(\xi) \simeq \textcircled{n_1(\xi)}\, N(\xi)$ and $T(\xi) \simeq \textcircled{n_2(\xi)}\, N(\xi)$ (theorem 1.43). Similarly $S(\xi) \simeq \textcircled{m_1(\xi)}\, R(\xi)$ and $T(\xi) \simeq \textcircled{m_2(\xi)}\, R(\xi)$ for μ_{II}-almost all ξ, again by theorem 1.43. Here $\xi \to n_i(\xi)$ $(i = 1,2)$ is an integral (or $+\infty$) valued Borel function (lemma 4.3) and $\xi \to m_i(\xi)$ $(i = 1,2)$ is a positive real (or $+\infty$) valued Borel function (lemma 4.4).

First suppose $S \lesssim T$. Then by proposition 3.23 we have $S(\xi) \lesssim T(\xi)$ for μ-almost all ξ and hence $n_1(\xi) \leq n_2(\xi)$ for μ_I-almost all ξ and $m_1(\xi) \leq m_2(\xi)$ for μ_{II}-almost all ξ. Hence

$$f_S^R(\nu_I) = \nu_I\text{-ess. inf.}\left\{n_1(\xi)\right\} \leq \nu_I\text{-ess. inf.}\left\{n_2(\xi)\right\} = f_T^R(\nu_I).$$

Similarly $f_S^R(\nu_{II}) \leq f_T^R(\nu_{II})$. Recall that $f_S^R(\nu_{III}) = f_T^R(\nu_{III}) = +\infty$. Thus

$$f_S^R(\nu) = \min\left\{f_S^R(\nu_i) : i = I,\ II,\ III\right\} \leq \min\left\{f_T^R(\nu_i) : i = I,\ II,\ III\right\} = f_T^R(\nu) .$$

Conversely suppose $f_S^R(\nu) \leq f_T^R(\nu)$ for all ν, $\nu << \mu$. Since m_1 and m_2 are μ_{II}-measurable functions on $\tilde{\mathfrak{F}}$ (lemma 4.4) we have, for each positive integer n, that

$$B_n = \left\{\xi : m_1(\xi) \geq m_2(\xi) + \frac{1}{n}\right\}$$

is a μ_{II}-measurable set. Suppose

$$B = \bigcup_{n=1}^{\infty} B_n = \left\{\xi : m_1(\xi) > m_2(\xi)\right\}$$

has positive μ_{II}-measure. Then there exists a positive integer n such that $\mu_{II}(B_n) \neq 0$. Let ν denote the restriction of μ_{II} to B_n, i.e., $\nu(C) = \mu_{II}(C \cap B_n)$ for every Borel subset C of $\tilde{\mathfrak{F}}$. Then we have $\nu << \mu_{II} << \mu$, $\nu \neq 0$, and

$$\begin{aligned} f_S^R(\nu) &= \nu\text{-ess. inf.}\left\{m_1(\xi)\right\} \\ &\geq \nu\text{-ess. inf.}\left\{m_2(\xi) + \frac{1}{n}\right\} \\ &\geq \nu\text{-ess. inf.}\left\{m_2(\xi)\right\} + \frac{1}{n} \\ &> \nu\text{-ess. inf.}\left\{m_2(\xi)\right\} = f_T^R(\nu) . \end{aligned}$$

Since this contradicts our hypothesis that $f_S^R(\nu) \leq f_T^R(\nu)$ for all $\nu << \mu$, we conclude that $\mu_{II}(B) = 0$ and hence $m_1(\xi) \leq m_2(\xi)$ for μ_{II}-almost all ξ. Similarly we may prove $n_1(\xi) \leq n_2(\xi)$ for μ_I-almost all ξ. Of course we have $S(\xi) \simeq T(\xi)$ for μ_{III}-almost all ξ (proposition 1.20). Hence we have $S(\xi) \lesssim T(\xi)$ for μ-almost all ξ in $\tilde{\mathfrak{F}}$. Hence by proposition 3.23 we have $S \lesssim T$.

COROLLARY 4.9. Let S and T denote quasi-equivalent operators and let μ denote their common central measure class on $\tilde{\mathfrak{F}}$. Let R denote any finite operator quasi-equivalent to the type II part of S. Then $S \simeq T$ if and only if $f_S^R(\nu) = f_T^R(\nu)$ for all measure classes ν, $\nu << \mu$.

We have shown that relative multiplicity functions distinguish operators up to unitary equivalence within a quasi-equivalence class. We next establish that every such relative multiplicity function does correspond to some unitary equivalence class.

THEOREM 4.10. Let t denote a quasi-equivalence class of operators and let μ denote the corresponding central measure class on $\tilde{\mathfrak{F}}$. Let $\mu = \mu_I \vee \mu_{II} \vee \mu_{III}$ be its unique decomposition into type I, II and III parts (cf. proposition 2.43). Let f denote a multiplicity function for μ (definition 4.1), such that it is integral (or $+\infty$) valued for any measure class ν such that $\nu << \mu_I$ and such that $f(\mu_{III}) = +\infty$. Further let S denote any finite operator in the type II part of t. Then there exists an operator T in the class t such that f is equal to the relative multiplicity function f_T^S associated with T, with respect to S (definition 4.5).

PROOF. We shall construct three operators, T_I, T_{II} and T_{III} such that $T = T_I \oplus T_{II} \oplus T_{III}$ has the desired properties. First let T_{III} be any operator in the type III part of t. (If the type III part is empty, simply omit T_{III} in the above direct sum.)

We next construct the operator T_I, assuming the type I part of t is non-empty (otherwise we simply omit the term T_I in the direct sum definition of T). Since μ_I is a type I central measure class, it corresponds to a type I quasi-equivalence class t_I, which, by the definition of type I (i.e., discrete), contains a multiplicity free operator, which we denote by M. Hence M has a central decomposition of the form

$$M \simeq \int_{\tilde{\mathfrak{F}}}^{\oplus} M(\xi)\, d\mu_I(\xi) .$$

By part 5 of theorem 3.4, μ_I - almost all the components $M(\xi)$ are irreducible operators. For each n, a positive integer or $+\infty$, let

$$\nu_n = \operatorname{Sup}\left\{ \nu : \nu << \mu_I \text{ and } f(\nu) \geq n \right\} .$$

(Note that since the central measure lattice on $\tilde{\mathfrak{F}}$ is a lattice ideal (corollary 3.11) it is boundedly complete.) By theorem 2, page 78 of [77] each $\nu_n << \mu$ is equivalent to a measure of the form μ_{B_n} for some Borel subset B_n of $\tilde{\mathfrak{F}}$, where $\mu_{B_n}(C) = \mu(B_n \cap C)$ for all Borel subsets C of $\tilde{\mathfrak{F}}$. Further if $n > m$ then $\nu_n << \nu_m$ and hence $\mu_{B_n} << \mu_{B_m}$. Theorem 1, page 78 of [77] implies that the Borel sets B_n have the property that $\mu\left(B_n - B_m\right) = 0$ whenever $n > m$. Thus without loss of generality we may choose the sequence of Borel sets B_n such that $B_n \subset B_m$ whenever $n > m$. Then for each n, a positive integer or $+\infty$, define

$$B_n' = B_n - \left(\bigcup_{m>n} B_m \right) .$$

Then

$$B_1 = \bigcup_{1 \leq n \leq +\infty} B_n' \quad \text{and} \quad \mu_I(\tilde{\mathfrak{F}} - B_1) = 0.$$

Since the map $\xi \to M(\xi)$ is a μ_I - essentially bounded Borel field of irreducible operators on B_1 we have for each n, a positive integer or $+\infty$, that $\xi \to \textcircled{n}M(\xi)$ is a μ_I-essentially bounded Borel field of factor operators on B_1 and the central decomposition of $\textcircled{n}M$ is given by

$$\textcircled{n}M \simeq \int_{\tilde{\mathfrak{F}}}^{\oplus} \textcircled{n}M(\xi)\, d\mu_I(\xi)$$

(cf. corollary 3.22). Then we define the operator T_I by

$$T_I = \sum_{1 \leq n \leq +\infty} \oplus \int_{B_n'}^{\oplus} \textcircled{n}M(\xi)\, d\mu_I(\xi)$$

and T_I is quasi-equivalent to M (use proposition 1.11 for example) and hence T_I is contained in the type I part t_I of t and furthermore $f_{T_I}(\nu) = f(\nu)$ for all $\nu << \mu_I$.

We next begin the process of constructing the operator T_{II}, assuming that the type II part of t is nonempty (otherwise we simply omit the term T_{II} in the definition $T = T_1 \oplus T_{II} \oplus T_{III}$).

For each λ in $\mathbb{R}^+ \cup \{+\infty\}$ define the measure class

$$\nu_\lambda = \mathrm{Sup}\left\{\nu : \nu << \mu_{II} \text{ and } f(\nu) \geq \lambda\right\}.$$

Since $\nu_\lambda << \mu$, theorem 2, page 78 of [77] implies that there exists a Borel set B_λ for each λ such that $\mu_{B_\lambda} \sim \nu_\lambda$ where $\mu_{B_\lambda}(C) = \mu(B_\lambda \cap C)$ for each Borel subset C of $\tilde{\mathfrak{F}}$. Since $\lambda_1 > \lambda_2$ implies $\nu_{\lambda_1} << \nu_{\lambda_2}$ we may (cf. theorem 1, page 78 of [77]), without loss of generality, choose the Borel sets B_λ such that

(1) $$B_{\lambda_1} \subset B_{\lambda_2} \quad \text{whenever } \lambda_1 > \lambda_2.$$

Note that since μ_{B_λ} is the supremum of measure classes ν for which $f(\nu) \geq \lambda$ it follows from condition 3 of definition 4.1 (cf. also remark 4.2) that

(2) $$f\left(\mu_{B_\lambda}\right) \geq \lambda, \qquad 0 < \lambda \leq +\infty .$$

We next show that, without loss of generality, the family of Borel sets B_λ may be chosen so as to satisfy the following continuity property:

(3) $$B_\lambda = \bigcap_{\alpha<\lambda} B_\alpha, \qquad 0 < \lambda \leq +\infty .$$

Since we already have that $B_\lambda \subset B_\alpha$ whenever $\alpha < \lambda$ we have

$$B_\lambda \subset \bigcap_{\alpha<\lambda} B_\alpha .$$

We next show that $\mu\left(\bigcap_{\alpha<\lambda} B_\alpha - B_\lambda\right) = 0$ and hence that we may adjust our definition of B_λ (by perhaps tacking on a Borel set of μ-measure 0) to obtain the desired property (3). Let $C_\lambda = \bigcap_{\alpha<\lambda} B_\alpha$. Then $C_\lambda \subset B_\alpha$ for all $\alpha < \lambda$ implies that $\mu_{B_\alpha} << \mu_{C_\lambda}$ for all $\alpha < \lambda$ which implies $f\left(\mu_{C_\lambda}\right) \geq f\left(\mu_{B_\alpha}\right) \geq \alpha$ for all $\alpha < \lambda$ by axiom 2 of definition 4.1 and property (2) established above. Hence $f\left(\mu_{C_\lambda}\right) \geq \lambda$. Since $\nu_\lambda \sim \mu_{B_\lambda}$ is defined to be the supremum of measures ν for which $f(\nu) \geq \lambda$ we must have $\mu_{C_\lambda} << \mu_{B_\lambda}$. On the other hand $B_\lambda \subset C_\lambda$ implies $\mu_{B_\lambda} << \mu_{C_\lambda}$ and hence $\mu_{C_\lambda} \sim \mu_{B_\lambda}$. Thus $\mu\left(C_\lambda - B_\lambda\right) = \mu\left(\bigcap_{\alpha<\lambda} B_\alpha - B_\lambda\right) = 0$.

Next for each λ, $0 < \lambda \leq +\infty$, define

(4) $$B'_\lambda = B_\lambda - \bigcup_{r>\lambda} B_r = B_\lambda - \bigcup_{\alpha>\lambda} B_\alpha$$

where the first union is taken over all rational numbers r greater than λ, assuring that each B'_λ is a Borel subset of $\tilde{\mathfrak{F}}$. Then a trivial verification will show that if $\lambda_1 \neq \lambda_2$, then

(5) $$B'_{\lambda_1} \cap B'_{\lambda_2} = \emptyset .$$

We next verify that

(6) $$\bigcup_{\alpha \geq \lambda} B'_\alpha = \bigcup_{\alpha \geq \lambda} B_\alpha = B_\lambda \quad \text{for} \quad 0 < \lambda \leq +\infty .$$

Indeed since $B'_\alpha \subset B_\alpha$ for each α we have $\bigcup_{\alpha \geq \lambda} B'_\alpha \subset \bigcup_{\alpha \geq \lambda} B_\alpha$. To verify containment in the other direction, let $\xi \in \bigcup_{\alpha \geq \lambda} B_\alpha$. Then the set $\{\alpha : \alpha \geq \lambda \text{ and } \xi \in B_\alpha\}$ is nonempty and we may define

$$\beta = \operatorname{Sup}\{\alpha : \alpha \geq \lambda \quad \text{and} \quad \xi \in B_\alpha\} .$$

Then $\lambda \leq \beta \leq +\infty$ and by our continuity property (3) we have

$$B_\beta = \bigcap_{\alpha < \beta} B_\alpha .$$

By our nesting property (1) above we have $\xi \in B_\alpha$ for all $\alpha < \beta$ and hence $\xi \in B_\beta$. Clearly $\xi \notin B_\alpha$ for any $\alpha > \beta$ by our definition of β. Hence $\xi \in B'_\beta = B_\beta - \bigcap_{\alpha > \beta} B_\alpha$. Since $\beta \geq \lambda$ we have $\xi \in \bigcup_{\alpha \geq \lambda} B'_\alpha$ and we have established (6).

We next verify

(7) $$\mu_{II}\Big(\tilde{\mathfrak{F}} - \bigcup_{\alpha > 0} B'_\alpha\Big) = \mu_{II}\Big(\tilde{\mathfrak{F}} - \bigcup_{\alpha > 0} B_\alpha\Big) = 0.$$

For this purpose we introduce the simplifying notation $B_0 = \tilde{\mathfrak{F}} - \bigcup_{\alpha > 0} B'_\alpha = \tilde{\mathfrak{F}} - \bigcup_{\alpha > 0} B_\alpha$ and $\eta = \mu_{II}$. Then $\eta_{B_0} << \eta << \mu$ and hence the multiplicity function f is defined at η_{B_0}. We first show that $f(\eta_{B_0}) \neq 0$. Indeed if $f(\eta_{B_0}) = 0$, then axiom 4 of definition 4.1 implies that there exists a nonzero measure ν, $\nu << \eta_{B_0}$ such that $f(\nu) = \lambda_1 > 0$. By the definition of B_{λ_1} we have $\nu << \eta_{B_{\lambda_1}}$. But since $B_0 \cap B_{\lambda_1} = \emptyset$ we also have that $\eta_{B_0} \perp \eta_{B_{\lambda_1}}$ which contradicts the fact that ν is nonzero. Hence we have

$f\left(\eta_{B_o}\right) \neq 0$. Let $f\left(\eta_{B_o}\right) = \lambda_o$ where $0 < \lambda_o \leq +\infty$. Then $\eta_{B_{\lambda_o}} = \operatorname{Sup}\left\{\nu : \nu << \eta \text{ and } f(\nu) \geq \lambda_o\right\}$. Thus $\eta_{B_o} << \eta_{B_{\lambda_o}}$ which implies $\eta\left(B_o - B_{\lambda_o}\right) = 0$. But $B_o \cap B_{\lambda_o} = \emptyset$ and hence $\eta\left(B_o\right) = 0$ which is precisely property (7) which we wished to establish.

We next define a positive real (and $+\infty$) valued function m on $\bigcup_{0<\alpha\leq+\infty} B'_\alpha$ by

$$m(\xi) = \lambda \quad \text{if and only if} \quad \xi \in B'_\lambda .$$

By property (5) above m is well-defined and by property (7), m is defined for μ_{II}-almost all ξ in $\tilde{\mathfrak{Z}}$. Further m is a Borel function. Indeed by (6) above we have

$$\left\{\xi : m(\xi) \geq \lambda\right\} = \bigcup_{\alpha\geq\lambda} B'_\alpha = B_\lambda , \tag{8}$$

which is a Borel set.

Next define the positive real (and $+\infty$) valued function g on the lattice ideal of central measure classes ν absolutely continuous with respect to μ_{II} as follows:

$$g(\nu) = \nu\text{-ess. inf.}\left\{m(\xi)\right\} \quad \text{for} \quad \nu << \mu_{II} .$$

(For $\nu = 0$ let $g(\nu) = +\infty$.)

We next verify that $g = f$ by proving, for any λ, $0 < \lambda \leq +\infty$, that $g(\nu) \geq \lambda$ if and only if $f(\nu) \geq \lambda$.

Suppose first that $f(\nu) \geq \lambda$. Then $\nu \sim \mu_B$ with $B \subset B_\lambda$ since $\nu << \mu_{B_\lambda}$ by the definition of B_λ. Since $m(\xi) \geq \lambda$ for all ξ in B_λ we have $m(\xi) \geq \lambda$ for all ξ in B. Hence

$$g(\nu) = \mu_B\text{-ess. inf.}\left\{m(\xi)\right\} \geq \lambda .$$

Conversely suppose $g(\nu) \geq \lambda$ where $\nu << \mu_{II}$. Then there exists a Borel set B such that $\nu \sim \mu_B$ and μ_B - ess. $\inf\left\{m(\xi)\right\} \geq \lambda$. Thus

$$\mu\left(\left\{\xi : m(\xi) < \lambda\right\} \cap B\right) = 0.$$

Further by condition (8) above we have

$$\left\{\xi : m(\xi) < \lambda\right\} = \left\{\xi : m(\xi) \geq \lambda\right\}^c = B_\lambda^c .$$

Thus $\mu\left(B_\lambda^c \cap B\right) = \mu\left(B - B_\lambda\right) = 0$. Hence $\mu_B << \mu_{B_\lambda}$ and thus by condition (2) above we have

$$f(\nu) = f\left(\mu_B\right) \geq f\left(\mu_{B_\lambda}\right) \geq \lambda .$$

Thus we have verified that $f = g$, i.e.,

(9) $$f(\nu) = \nu\text{- ess. inf.}\left\{m(\xi)\right\}, \qquad \nu << \mu_{II}.$$

We next turn our attention to the specific construction of the operator T_{II}. For this purpose we first note that every positive real number λ has a canonical dyadic expansion of the form

$$\lambda = n(\lambda) + \sum_{i=1}^{+\infty} \delta_i(\lambda)\, 2^{-i}$$

defined as follows. For each positive real λ, $n(\lambda)$ is the largest integer less than λ. The functions δ_i are defined inductively as follows:

$$\begin{aligned} \delta_1(\lambda) &= 1 \quad \text{if} \quad n(\lambda) + \tfrac{1}{2} < \lambda \\ &= 0 \quad \text{otherwise} \end{aligned}$$

and assuming $\delta_1, \ldots, \delta_{k-1}$ have been defined, we define

$$\delta_k(\lambda) = 1 \quad \text{if} \quad n(\lambda) + \sum_{i=1}^{k-1} \delta_i(\lambda)2^{-i} + 2^{-k} < \lambda$$

$$= 0 \quad \text{otherwise.}$$

It is convenient to trivially extend this dyadic expansion formula to the case $\lambda = +\infty$ by defining $n(+\infty) = +\infty$ and $\delta_i(+\infty) = 0$ for $i = 1,2,\ldots$. Then it is clear that each δ_i is a Borel function from the extended positive real axis $(0, +\infty]$ into the two point set $\{0,1\}$.

We next take the dyadic expansion of $m(\xi)$, for ξ in $\tilde{\mathfrak{F}}$, where $\xi \to m(\xi)$ is the Borel function defined earlier. Then

$$m(\xi) = n\Big(m(\xi)\Big) + \sum_{i=1}^{\infty} \delta_i\Big(m(\xi)\Big)2^{-i}.$$

Now S is the given finite operator in the type II part of t. Let

$$S \simeq \int_{\tilde{\mathfrak{F}}}^{\oplus} S(\xi)\, d\mu_{II}(\xi)$$

denote the central decomposition of S and let B denote a Borel subset of $\tilde{\mathfrak{F}}$ whose relative complement in $\tilde{\mathfrak{F}}$ is a μ_{II}-null set such that, for each $\xi \in B$, $m(\xi)$ is defined and $S(\xi)$ is a finite type II factor operator, $\|S(\xi)\| \leq \|S\|$ (cf. parts 1 and 6 of theorem 3.4). Now clearly $\lambda \to n(\lambda)$ is a Borel function, as is $\xi \to m(\xi)$ and hence $\xi \to n(m(\xi))$ is a Borel function on B. For each nonnegative integer (or $+\infty$) p, let

$$B_p = \Big\{\xi : \xi \in B \text{ and } n\Big(m(\xi)\Big) = p\Big\}.$$

Then $B = \bigcup_{0 \leq p \leq +\infty} B_p$ and on each Borel set B_p

$$\xi \to \text{\textcircled{$n(m(\xi))$}}\, S(\xi) = \text{\textcircled{p}}\, S(\xi)$$

is a Borel field of operators by corollary 3.22. (In the case $p = 0$ one may think of $\textcircled{0}\, S(\xi)$ as a trivial operator acting on a zero dimensional space, i.e., in the direct sum definition to follow, such terms just don't appear.)

Further for each i, $i = 1, 2, \ldots$, $\left\{\xi : \delta_i\left(m(\xi)\right) = 0\right\}$ is a Borel subset of B and hence for each i

$$\xi \to \textcircled{\delta_i(m(\xi))}\, S(\xi)$$

is a Borel field of operators which is just $\xi \to S(\xi)$ on the Borel set $\left\{\xi : \delta_i\left(m(\xi)\right) = 1\right\}$ and is the trivial map ξ goes into the 0 operator on a 0-dimensional space on the Borel set $\left\{\xi : \delta_i\left(m(\xi)\right) = 0\right\}$. Thus

$$\xi \to \textcircled{2^{-i}}\,\textcircled{\delta_i(m(\xi))}\, S(\xi)$$

is a Borel field of operators on B by corollary 3.25 (with $\lambda = 2^{-i}$). Hence

$$\xi \to \textcircled{n(m(\xi))}\, S(\xi) \;\oplus\; \sum_{i=1}^{\infty}{}^{\oplus}\; \textcircled{2^{-i}\,\delta_i(m(\xi))}\, S(\xi)$$

$$\simeq \textcircled{n(m(\xi)) + \sum_{i=1}^{\infty} 2^{-i}\,\delta_i(m(\xi))}\, S(\xi) = \textcircled{m(\xi)}\, S(\xi)$$

is a Borel field of operators on B, by lemma 3.19. Further by lemma 3.21, and corollaries 3.22 and 3.25 we have that

$$T_{II} = \int_{\tilde{\mathfrak{F}}}^{\oplus} \textcircled{m(\xi)}\, S(\xi)\, d\mu_{II}(\xi)$$

is a type II operator quasi-equivalent to S and the above decomposition is the central decomposition of T_{II}.

Thus by definition one has

$$f^S_{T_{II}}(\nu) = \nu\text{-ess. inf.}\left\{m(\xi)\right\}$$

for all $\nu << \mu_{II}$, $\nu \neq 0$. Hence by property (9) above

$$f^S_{T_{II}}(\nu) = f(\nu) \quad \text{for all} \quad \nu, \quad \nu << \mu_{II}.$$

Putting all three constructions together we define $T = T_I \oplus T_{II} \oplus T_{III}$ and then we have, for $\nu << \mu_{II}$

$$\begin{aligned} f^S_T(\nu) &= \min\left\{f^S_T(\nu_i) : i = I, II, III\right\} \\ &= \min\left\{f(\nu_i) : i = I, II, III\right\} = f(\nu). \end{aligned}$$

REMARK 4.11. We conclude this section with a short digression on representation theory. In section 6 of [50] the author developed an analogous multiplicity theory for unitary representations of separable locally compact groups as well as for $*$-representations of separable C^*-algebras. The theory there is deficient in one respect. It does not contain a theorem like the one (theorem 4.10) which we just proved, i.e., it does not establish that to each multiplicity function that there is a corresponding unitary equivalence class of representations. However, the above theorem and its proof may be adapted to provide the missing theorem in [50]. Specifically our definition of multiplicity function (definition 4.1), particularly the addition of axiom 4, must be adapted to the representation theoretic context. When that is done these techniques give the following theorem, which now completes the theory of section 6 of [50].

THEOREM 4.12. Let G be a separable locally compact group and let L denote a quasi-equivalence class of separable continuous unitary representations of G. Let μ denote the corresponding canonical (central) measure class on the quasi-dual $\tilde{G}$ of G. Let f denote a multiplicity function defined on the measure classes ν, $\nu << \mu$ such that it is positive integral (or $+\infty$) valued for any ν, $\nu << \mu_I$, where μ_I is the type I part of μ and $f(\nu) = +\infty$ for any $\nu << \mu_{III}$, the type III part of μ. Further let M denote a finite representation contained in the type II part of L. Then there exists a representation $\bar{L}$ of G, $\bar{L} \in L$ such that

$$f(\nu) = f_{\mu,M}(\bar{L}, \nu)$$

for all $\nu << \mu$, where $f_{\mu,M}(\bar{L}, \cdot)$ denotes the multiplicity function defined on page 272 of [50].

A similar theorem holds for separable $*$-representations of a separable C^*-algebra.

Chapter 5

APPLICATIONS, EXAMPLES AND OPEN QUESTIONS

1. Smooth operators

We have already indicated many times throughout this memoir the rather sharp phenomenological boundary that exists between smooth and nonsmooth operators (cf. propositions 1.51, 1.54, 2.44, 2.56, 3.13, 3.17 and theorem 2.32). It is not surprising therefore that this entire theory becomes vastly simpler and more utilitarian and esthetic, when we restrict our attention to smooth operators. Further the analogy with the classical theory for normal operators is quite remarkable in the smooth case. It is only to this smooth (i.e., GCR) case that William Arveson refers in section 2.3 of [6] and the last paragraph of [3].

First of all if T is a smooth operator we need not even consider the concept of quasi-spectrum $\tilde{T}$ since $\hat{T} = \tilde{T}$ (theorem 2.32). Thus we may express the central decomposition in terms of irreducible operators rather than factor operators, as we shall make clear in a moment. Further none of the nasty technical complications associated with intractable Borel structures arise as $\hat{T}$ is smooth, i.e., is a standard Borel space. Thus the unitary equivalence classes of irreducible operators weakly contained in T can be distinguished by a countable number of Borel invariants. Hence for a smooth operator T we can reasonably aspire to the explicit computation and exhibition of the spectrum $\hat{T}$ of T. Another great blessing we obtain when we restrict our attention to smooth operators T is that we can forget one of the most technical parts of the theory, section 2 of Chapter 3, as every standard measure class on $\hat{T}$ is a central measure class.

We next take a closer look at the central decomposition of a smooth operator T,

$$T \simeq \int_{\hat{T}}^{\oplus} T(\xi)\, d\mu(\xi).$$

Here each $T(\xi)$ is a type I factor operator and hence of the form $T(\xi) \simeq \textcircled{n(\xi)} S(\xi)$ where $n(\xi)$ is a positive integer or $+\infty$, and $S(\xi)$ is an irreducible operator contained in the unitary equivalence class ξ (parts 1 and 6 of theorem 3.4 and corollary 1.44). By lemma 4.3 we have $\xi \to n(\xi)$ is a Borel function on $\hat{T}$. Thus for $1 \leq n \leq +\infty$ we may define the Borel set

$$\hat{T}(n) = \left\{ \xi : \xi \in \hat{T} \quad \text{and} \quad n(\xi) = n \right\}$$

which gives a Borel partition of $\hat{T}$,

$$\hat{T} = \bigcup \left\{ \hat{T}(n) : 1 \leq n \leq +\infty \right\}.$$

Hence

$$T \simeq \sum_{1 \leq n \leq +\infty} \oplus \int_{\hat{T}(n)}^{\oplus} T(\xi)\, d\mu(\xi)$$

$$\simeq \sum_{1 \leq n \leq +\infty} \oplus \int_{\hat{T}(n)}^{\oplus} \textcircled{n} S(\xi)\, d\mu(\xi)$$

since $n(\xi) = n$ for all ξ in $\hat{T}(n)$. Now

$$M_n = \int_{\hat{T}(n)}^{\oplus} S(\xi)\, d\mu(\xi)$$

is a multiplicity free operator (cf. part 5 of theorem 3.4) and

$$\textcircled{n} M_n = \textcircled{n} \int_{\hat{T}(n)}^{\oplus} S(\xi)\, d\mu(\xi) \simeq \int_{\hat{T}(n)}^{\oplus} \textcircled{n} S(\xi)\, d\mu(\xi)$$

is a homogeneous operator of order n (cf. definition 1.21 and corollary 3.22).

Thus a slight modification of the central decomposition of T produces a decomposition of T into irreducible components rather than factor components

$$(1) \qquad T \simeq \sum_{1 \le n \le +\infty} \oplus \, \textcircled{n} \int_{\hat{T}(n)}^{\oplus} S(\xi)\, d\mu(\xi).$$

This decomposition into irreducible operators is equivalent to that achieved by taking the central decomposition of each of the multiplicity free operators appearing in the decomposition given by proposition 1.27.

Proposition 1.27 can also be derived from the usual theory of multiplicity functions [77]. A measure μ is said to be uniformly of multiplicity n relative to a multiplicity function f which has μ in its domain if $f(\nu) = n$ for all ν, $\nu \neq 0$, and $\nu << \mu$ (cf. page 81 of [77]). Thus a type I operator T is homogeneous of order n (definition 1.21) if and only if its corresponding central measure μ on $\hat{T}$ is uniformly of multiplicity n relative to the multiplicity function f associated with T. Hence the decomposition of a type I operator as a direct sum of homogeneous operators (proposition 1.27) follows from the spectral multiplicity theory. Indeed since in the type I case our multiplicity function is positive integral (or $+\infty$) valued, the result follows from theorem 3 page 82 of [77] which asserts that $\mu = \vee_j \mu_j$ where $\{\mu_j\}$ is a countable orthogonal family of measure classes such that, for each j, either $\mu_j = 0$ or μ_j has uniform multiplicity j relative to the multiplicity function f, where $1 \leq j \leq +\infty$.

One of the remarkable features of smooth operators is the extent to which the spectral multiplicity theory takes precisely the same form as the classical theory for normal operators. Since the quasi-equivalence class of a smooth operator never has a type II part, there is no need for the notion of a <u>relative</u> multiplicity function and instead we have a well defined positive integral (or $+\infty$) valued multiplicity function associated with each unitary equivalence class of operators quasi-equivalent to a smooth operator T. The multiplicity functions catalog all the unitary equivalence classes of operators quasi-equivalent to T and are defined on the ideal of all standard Borel measures on $\hat{T}$ absolutely continuous with respect to the central measure class of T on $\hat{T}$. In this case the definition of multiplicity function is almost precisely the same as that used in the spectral multiplicity theory for normal operators (page 80 of [77]). Indeed for the case of smooth operators (and hence for cardinal valued multiplicity functions which never take on the value 0) axiom 4 of definition 4.1 is vacuous and thus dispensable. Thus the only difference remaining between our definition of multiplicity function (definition 4.1) and the one given by Paul Halmos [77], is a minor distinction involving personal taste and convenience. If ν_o denotes the trivial measure on the spectrum $\hat{T}$ (i.e., $\nu_o(\hat{T}) = 0$) we have required that a multiplicity function f have the value $+\infty$ at ν_o while Halmos requires that $f(\nu_o) = 0$.

Actually all the remarks of the preceding paragraph on the similarity of the generalized spectral multiplicity theory to the classical theory apply equally well to a much larger class of operators, namely the type I operators (cf. part 6 of theorem 3.4). Further the decomposition (1) into irreducible operators discribed earlier is also valid for type I operators. The decomposition (1) is essentially one form of the central decomposition of T and if T is type I but not smooth (such animals exist – proposition 1.52) it is known that there are still other (non-central) ways of decomposing T as a direct integral of irreducible operators. This pathological property of non-smooth operators has not plagued this memoir as we have based the entire theory on the essentially unique (proposition 3.2) and salubrious (theorem 3.4) central decomposition of any operator into factor operators.

2. Normal operators and spectral measures.

In the classical case of the spectral theorem of a normal operator one defines a spectral measure E on a Borel space S to be a projection valued measure on S. That is to say, for each Borel subset B of S, $E(B)$ is a projection on a Hilbert space $\mathcal{K}$, $E(S) = I$, the identity operator on $\mathcal{K}$ and E is countably additive, i.e., if $\{B_n\}$ is a sequence of mutually disjoint Borel subsets of S, then

$$E\left(\cup_n B_n\right) = \Sigma_n E\left(B_n\right)$$

(cf. page 58 of [77]).

Associated with any direct integral decomposition of an operator T, say

$$T = \int_S^{\oplus} T(\xi)\, d\mu(\xi)$$

there is associated a natural spectral measure, namely that defined on S by

$$E(B) = \int_B^{\oplus} I(\xi)\, d\mu(\xi)$$

where $\xi \to I(\xi)$ is the Borel map defined on S such that $I(\xi)$ is the identity operator on $\mathcal{K}\big(T(\xi)\big)$ for μ - almost all ξ and B is a Borel subset of S. We can characterize the notion of central decomposition in terms of its associated spectral measure.

PROPOSITION 5.1. A direct integral decomposition

$$T = \int_S^{\oplus} T(\xi)\, d\mu(\xi)$$

is central if and only if the range of the associated spectral measure is precisely the set of projections in the center of $\mathcal{A}(T)$.

PROOF. Since each projection, say $E(B)$, in the range of the spectral measure corresponds to the characteristic function of the set B, the range of the spectral measure is precisely the set of projections in the abelian von Neumann algebra of all diagonalizable operators of the decomposition. Recall that every von Neumann algebra is generated by its projections (corollary 2, page 4 of [37]). Thus the range of the associated spectral measure is the set of projections in the center of $\mathcal{A}(T)$ if and only if the set of diagonalizable operators is the center of $\mathcal{A}(T)$, i.e., if and only if the decomposition is central (definition 3.3).

Let T be a normal operator. We will now show that the central decomposition of T is just the usual spectral theorem and that the associated spectral measure is precisely the spectral measure of that classical theorem (theorem 1, page 71 of [77]). We have already noted that if T is normal then it is smooth so that $\hat{T} = \tilde{T}$ and in fact the spectrum $\hat{T}$ of T may be identified with the classical numerical spectrum $\Lambda(T)$ (cf. remarks preceding proposition 2.49 and following proposition 2.56). This identification of $\hat{T}$ with $\Lambda(T)$ is clear as every irreducible operator weakly contained in a normal operator is one dimensional and hence may be identified with a complex number. Consider the central decomposition of T,

$$T = \int_{\hat{T}=\Lambda(T)}^{\oplus} T(\xi)\, d\mu(\xi)$$

and let E denote the corresponding spectral measure defined on $\Lambda(T)$. For each $\xi \in \Lambda(T)$, $T(\xi)$ is quasi-equivalent to the one-dimensional operator ξ

and hence (theorem 1.43) is necessarily of the form

$$T(\xi) = \xi I(\xi)$$

for μ-almost all ξ, where $I(\xi)$ is the identity operator on $\mathcal{H}(T(\xi))$. Thus the central decomposition of T has the form

$$T = \int_{\Lambda(T)}^{\oplus} \xi I(\xi)\, d\mu(\xi)$$

$$= \int_{\Lambda(T)} \xi dE(\xi) .$$

Thus we see that the central decomposition of a normal operator is in fact equivalent to the classical spectral theorem in which the operator is represented by a spectral (i.e., projection valued) measure. Nevertheless the central decomposition theory casts the classical spectral theory into a slightly different mold and this new perspective on the spectral theorem is itself instructive. While the notion of a spectral measure may well appeal esoteric to the novice, the central decomposition form of the spectral theorem, i.e.,

$$T \simeq \int_{\Lambda(T)}^{\oplus} \xi I(\xi)\, d\mu(\xi)$$

is a natural and intuitive generalization of the finite dimensional case, where a normal matrix is unitarily equivalent to a diagonal matrix. Indeed even in the von Neumann-Dixmier terminology (definition 3, page 162 of [36]) T is a diagonalizable operator relative to the above decomposition. From this point of view our (central) decomposition theory for arbitrary operators,

$$T \simeq \int_{\tilde{T}}^{\oplus} T(\xi)\, d\mu(\xi)$$

may be considered as a generalized spectral theorem.

We have already established in the previous section that our spectral multiplicity theory (chapter 4) takes the classical form for any smooth operator. Hence in particular it reduces to the classical theory (chapter 3 of [77]) in the normal case.

3. Binormal operators

Since we have just observed that an operator T is normal if and only if its spectrum $\hat{T}$ consists only of one-dimensional irreducible operators, it is natural to examine next the class of operators T for which the spectrum $\hat{T}$ consists of at most two-dimensional irreducible operators, i.e., those operators for which $\hat{T} = {}_2\hat{T}$, in the notation introduced following proposition 2.56. Such operators have been christened binormal operators by Arlen Brown [16]. Harry Gonshor [71] has developed and studied the spectral theorem for this class of operators, which he called J_2 operators.

In Gonshor's terminology [71] an operator is called a J_n operator (also called n-normal in [16] and [20]) if $\hat{T} = {}_n\hat{T}$, i.e., if all the irreducible operators weakly contained in T are at most n-dimensional.[1)]

---

1) Every scholarly memoir should have at least one technical footnote. This is ours. The definition of J_n-operator quoted above is not precisely that given by Gonshor in [71] but we now show that it is in fact equivalent to Gonshor's definition. Gonshor defined an operator to be J_n if it has a central decomposition into type I factor operators such that all the corresponding irreducible operators have dimension at most n (cf. theorem 1.43 of this memoir). If $\hat{T} = {}_n\hat{T}$ of course the above condition of Gonshor holds (theorem 3.5 and theorem 2.32). However suppose T satisfies Gonshor's condition, i.e., if μ is the central measure class on $\hat{T}$ determined by the central decomposition of T, then μ-almost all the elements of $\hat{T}$ have

The class of J_n operators is clearly invariant with respect to weak equivalence (proposition 2.35). If T is an operator and μ denotes the central measure class on its spectrum $\hat{T}$, then T is called a pure J_n operator (also called homogeneous n-normal operator) if $\hat{T} - \hat{T}_n$ is a μ-null set, where $\hat{T}_n$ denotes the n-dimensional part of $\hat{T}$. The class of pure J_n operators is not invariant with respect to weak equivalence but is invariant with respect to quasi-equivalence (theorem 3.8).

Harry Gonshor's spectral theorem for binormal operators (theorems 3, 4 and 6 of [71]) is precisely the central decomposition, where he has further indicated the precise form of the component irreducible operators. His spectral multiplicity theory, while different in certain respects from the multiplicity theory developed here (chapter 4) achieves the same purpose, namely solving the unitary equivalence problem for binormal operators. Further the reader should be warned that the notion of "abstract spectrum" introduced by Gonshor in [72] does not correspond to the notion of spectrum $\hat{T}$ introduced in this memoir.

The specific parametrization of the two-dimensional part $\hat{T}_2$ of the spectrum $\hat{T}$ of any operator T requires certain known results on the unitary equivalence problem for 2×2 complex matrices (cf. lemma 8.1 of [16] and

---

dimension at most n. Thus $\hat{T} - {}_n\hat{T}$ is a μ-null set. However $\hat{T}$ has a topology and relative to this topology ${}_n\hat{T}$ is a closed subset of $\hat{T}$ (cf. remarks following proposition 2.56). Further proposition 8.6.8 of [36] may be translated into the operator theoretic context to assert that the support of μ (i.e., the smallest closed subset of $\hat{T}$ whose complement is a μ-null set) is $\hat{T}$. Thus since $\hat{T} - {}_n\hat{T}$ is a μ-null set and ${}_n\hat{T}$ is closed we may conclude that $\hat{T} = {}_n\hat{T}$.

[13] and [97]). Every such unitary equivalence class admits a canonical form which is upper triangular, i.e.,

$$\begin{pmatrix} x & y \\ 0 & z \end{pmatrix} \tag{1}$$

where $y \geq 0$ and x and z are complex numbers. Since

$$\begin{pmatrix} x & y \\ 0 & z \end{pmatrix} \quad \text{and} \quad \begin{pmatrix} z & y \\ 0 & x \end{pmatrix}$$

are unitarily equivalent, it is necessary to make an additional (artificial) decision as to the order in which the complex eigenvalues are to be given. Rather arbitrarily we pronounce that the complex numbers $\alpha = \alpha_1 + i\alpha_2$ are to be given the lexicographical ordering associated with linear ordering of the reals and decree that the canonical form (1) will have the property that $x \leq z$. Further it is easy to see that the matrix (1) is irreducible if and only if $y \neq 0$.

By using this canonical form, the space of all unitary equivalence classes of 2×2 irreducible complex matrices may be parametrized (letting $x = r_1 + ir_2$, $z = r_3 + ir_4$, $y = r_5$) by the following Borel subset B of $\mathbb{R}^5$.

$$B = \left\{\left(r_1, r_2, r_3, r_4, r_5\right) : r_5 > 0 \text{ and either } r_1 < r_3 \text{ or } r_1 = r_3 \text{ and } r_2 \leq r_4\right\}$$

Of course all 2×2 complex matrices may be parametrized by $\mathbb{R}^8$ in the following manner

$$\begin{pmatrix} r_1 + ir_2 & r_5 + ir_6 \\ r_7 + ir_8 & r_3 + ir_4 \end{pmatrix}$$

in which case we may consider the set B as a Borel subset of R^8 ($r_6 = r_7 = r_8 = 0$). With this convention we see that B is a Borel cross-section of the space of unitary equivalence classes of 2×2 irreducible complex matrices. As a Borel subset of the standard Borel space R^8, B is a standard Borel space (corollary 1, page 139 of [96]). Further for any operator T, the set of those irreducible 2×2 complex matrices weakly contained in T is a Borel subset (of R^8) and hence also a standard Borel space. This follows from corollary 2.25 and the fact [126] that on the 2×2 matrices all the operator topologies are the same as the uniform topology and hence the Borel structure as we have defined it (proposition 2.3) is the same as the usual Borel structure in R^8. Thus for any operator T, the quotient map (with respect to unitary equivalence) which maps the standard Borel space $T^c \cap B$ onto the standard Borel space $\hat{T}_2$ is a one-to-one Borel map and hence by theorem 3.2 of [96] is a Borel isomorphism. We have thus established:

PROPOSITION 5.2. If T is any operator and if we parametrize the two-dimensional part $\hat{T}_2$ of the spectrum $\hat{T}$ of T, by means of the canonical form (1) as a Borel subset of R^5, then the Borel structure of $\hat{T}_2$ corresponds to the usual Borel structure associated with the usual topology of R^5.

The fact that the parametrization associated with the canonical form (1) gives rise to the correct Borel structure makes all matters relating to direct integrals and central measures on $\hat{T}_2$ much simpler. The one-dimensional part $\hat{T}_1$ (the normal spectrum) of the spectrum $\hat{T}$ of an operator

T may clearly be parametrized by $\mathbb{C}$ or $\mathbb{R}^2$. Hence if T is a binormal operator, there is a canonical way of parametrizing the spectrum $\hat{T}$ of T as a Borel subset of $\mathbb{R}^7$, such that the Borel structure of $\hat{T}$ corresponds to the usual Borel structure of $\mathbb{R}^7$. This fact contrasts sharply with the topological situation, which this parametrization does not respect at all. Indeed John Bunce and James Deddens [20] have explicitly computed the topology of the spectrum $\hat{T}$ for binormal operators T and in general it is not even Hausdorff. It seems that as far as the Borel structure is concerned, the one-dimensional and the two-dimensional parts of $\hat{T}$ are nicely separated (they are both Borel sets) but they are thoroughly entangled relative to the topology of $\hat{T}$, even though $\hat{T}_1$ is a closed subset of $\hat{T}$.

Let T be a binormal operator with spectrum $\hat{T} = {}_2\hat{T} = \hat{T}_1 \cup \hat{T}_2$ (cf. notation introduced following proposition 2.56). Preceding proposition 4, page 304 of [20], Bunce and Deddens quote a result of Arlen Brown [16] which asserts that T may be decomposed in the form

$$T = N \oplus \begin{pmatrix} X & Y \\ 0 & Z \end{pmatrix} \tag{2}$$

where N is a normal operator, X, Y and Z are commuting normal operators and Y is a positive injective operator. Bunce and Deddens use the decomposition (2) to describe $\hat{T}$ and its topology in terms of the spectrum of N and the joint spectrum of X, Y and Z.

We next indicate how the decomposition (2) may be obtained from the central decomposition of T and in so doing indicate how the spectra of N, X, Y and Z, and their spectral decompositions are related to the spectrum $\hat{T}$ of T and the central measure class μ on $\hat{T}$. Consider the central decomposition of T,

$$T \simeq \int_{\hat{T}}^{\oplus} T(\xi)\, d\mu(\xi) = \int_{\hat{T}_1}^{\oplus} T(\xi)\, d\mu(\xi) \oplus \int_{\hat{T}_2}^{\oplus} T(\xi)\, d\mu(\xi) .$$

Then

$$N = \int_{\hat{T}_1}^{\oplus} T(\xi)\, d\mu(\xi)$$

is a normal operator and the restriction of μ to the one-dimensional part $\hat{T}_1$ of $\hat{T}$ gives rise to the spectral theorem for N as described in the previous section, where $\hat{T}_1 = \Lambda(N)$, the ordinary spectrum of N, and $T(\xi) = \xi I(\xi)$ where $I(\xi)$ is an identity operator. The operator N is called the normal kernel by Brown [16].

We are thus left to consider the reduced binormal operator (or pure J_2 operator)

$$S = \int_{\hat{T}_2}^{\oplus} T(\xi)\, d\mu(\xi) .$$

Under the parametrization of $\hat{T}_2$ given by proposition 5.2, we may assume, without loss of generality, that each $T(\xi)$ has the form $m(\xi)$ $S(\xi)$ where $m(\xi)$ is a positive integer or $+\infty$ and

$$S(\xi) = \begin{pmatrix} x(\xi) & y(\xi) \\ 0 & z(\xi) \end{pmatrix}$$

where $y(\xi) > 0$ and $x(\xi) \leq z(\xi)$ relative to the lexicographical ordering of the complex numbers. Indeed if ξ corresponds to the vector r in R^5 under the parametrization given by proposition 5.2, then $x(\xi) = r_1 + ir_2$, $z(\xi) = r_3 + ir_4$ and $y(\xi) = r_5$. Thus $\xi \to x(\xi)$, $\xi \to z(\xi)$ are Borel maps of $\hat{T}_2$ into C and $\xi \to y(\xi)$ is a Borel map of $\hat{T}_2$ into R^+, and all three maps are bounded by $\|T\|$.

For each m, a positive integer or $+\infty$, let

$$\hat{T}_2(m) = \left\{ \xi : \xi \in \hat{T}_2 \quad \text{and} \quad m(\xi) = m \right\}.$$

Then $\left\{ \hat{T}_2(m) : 1 \leq m \leq +\infty \right\}$ is a Borel partition of $\hat{T}_2$ by lemma 4.3. For each positive integer m, let $\mathcal{H}_m$ denote the Hilbert space of m-tuples of complex numbers and let $\mathcal{H}_\infty = \ell_2$. Then we define multiplication operators X, Y, Z on

$$\mathcal{H} = \sum_m \oplus\, L^2\left(\hat{T}_2(m), \mu\, ; \mathcal{H}_m \right)$$

by

$$X\left\{ f_m(\xi) \right\} = \left\{ x(\xi)\, f_m(\xi) \right\}$$

$$Y\left\{ f_m(\xi) \right\} = \left\{ y(\xi)\, f_m(\xi) \right\}$$

$$Z\left\{ f_m(\xi) \right\} = \left\{ z(\xi)\, f_m(\xi) \right\}$$

where $f_m \in L^2\left(\hat{T}_2(m), \mu\, ; \mathcal{H}_m \right)$.

Then

$$\begin{pmatrix} X & Y \\ 0 & Z \end{pmatrix} \quad \text{acts on} \quad \mathcal{H} \oplus \mathcal{H}$$

and is unitarily equivalent to

$$S \simeq \sum_m \oplus\, \textcircled{m} \int^{\oplus}_{\hat{T}_2(m)} \begin{pmatrix} x(\xi) & y(\xi) \\ 0 & z(\xi) \end{pmatrix} d\mu(\xi)$$

under the natural identification of $\mathcal{H} \oplus \mathcal{H}$ with

$$\sum_m \oplus\, \textcircled{m} \left[L^2\left(\hat{T}_2(m), \mu \right) \oplus L^2\left(\hat{T}_2(m), \mu \right) \right].$$

We thus obtain the decomposition (2) and also see explicitly how the central decomposition of S is related to the definition of X, Y and Z as multiplication operators and hence to the spectral decompositions of X, Y and Z. Indeed let $\mathcal{S}(X)$ denote the image of $\hat{T}_2$ under the mapping of $\hat{T}_2$ into $\mathbb{C}$,

$$\begin{pmatrix} x & y \\ 0 & z \end{pmatrix} \to x \qquad \text{or} \qquad r \to r_1 + ir_2$$

and let μ_X denote the measure μ induces on $\mathcal{S}(X)$. Then $\mathcal{S}(X)$ differs from $\Lambda(X)$ by at most a μ_X-null set and μ_X denotes the measure class on $\Lambda(X)$ determined by the central decomposition (i.e., the spectral theorem) for X. Similar statements hold for the normal operators Y and Z.

4. Trinormal operators

An operator T is trinormal (or J_3 in the terminology of [71]) if $\hat{T} = {}_3\hat{T}$, i.e., if every irreducible operator weakly contained in T has dimension at most 3. In order to study the spectra of such operators in a very concrete way we must parametrize the space of unitary equivalence classes of 3×3 irreducible matrices. Hence we once again make use of the extensive work that has been done on the unitary equivalence problem for matrices (cf. [13], [97], [105], [112], [119]).

PROPOSITION 5.3. There exists a Borel cross-section of the space of all unitary equivalence classes of irreducible 3×3 complex matrices whereby each class has precisely one canonical representative in the upper-triangular form

$$\begin{pmatrix} a & d & f \\ 0 & b & g \\ 0 & 0 & c \end{pmatrix}$$

where

1. $a \le b \le c$ in the lexicographic ordering of the complex numbers
2. $d, g \ge 0$, but not both are zero
3. if either d or $g = 0$ then $f > 0$
4. if $a = b$ then $d > 0$
   if $b = c$ then $g > 0$.

PROOF. We begin with a basic upper triangular form described by Francis Murnaghan [97] and summarized by Carl Pearcy in lemmas 2.1, 2.2 and 2.3 of [105], which is a canonical form for 3×3 complex matrices with respect to unitary equivalence. We have performed the simple but tedious computations needed to determine which of these matrices are irreducible. Finally in describing the canonical form for these irreducible matrices, we have collapsed the three cases considered by Pearcy into a single account. We spare the reader the unenlightening details.

COROLLARY 5.4 If T is any operator, then the three dimensional part $\hat{T}_3$ of the spectrum $\hat{T}$ of T may be parametrized by a Borel subset of $\mathbb{R}^{10}$ (and the Borel structure of $\hat{T}_3$ then corresponds to the usual Borel structure of $\mathbb{R}^{10}$) by means of the canonical form given by proposition 5.3. Any vector r in $\mathbb{R}^{10}$ which corresponds to a point in $\hat{T}_3$ will satisfy the following six conditions, where r_i denotes the ith component of the vector r.

1. $r_1 < r_3$ or $r_1 = r_3$ and $r_2 \le r_4$

2. $r_3 < r_5$ or $r_3 = r_5$ and $r_4 \le r_6$

3. $r_7 \ge 0$ and $r_8 \ge 0$ but not both are zero

4. if $r_7 = 0$ or $r_8 = 0$ then $r_9 > 0$ and $r_{10} = 0$

5. if $r_1 = r_3$ and $r_2 = r_4$ then $r_7 > 0$

6. if $r_3 = r_5$ and $r_4 = r_6$ then $r_8 > 0$.

PROOF. We have considered the following specific parametrization induced by the canonical form given in proposition 5.3:

$$a = r_1 + ir_2, \qquad b = r_3 + ir_4, \qquad c = r_5 + ir_6$$

$$d = r_7 \qquad g = r_8 \qquad f = r_9 + ir_{10}.$$

The fact that the Borel structure is the correct one is obtained by the same argument we used to establish proposition 5.2.

Let T be a trinormal operator with spectrum $\hat{T} = {}_3\hat{T}$ and central measure class μ on $\hat{T}$. Then $\hat{T} = \hat{T}_1 \cup \hat{T}_2 \cup \hat{T}_3$ is a Borel partition of $\hat{T}$ and hence the central decomposition of T may be used to express T as a (disjoint) direct sum of a normal operator, a pure J_2 (or homogeneous binormal) operator and a pure J_3 (or homogeneous trinormal) operator as follows:

$$T \simeq \int_{\hat{T}}^{\oplus} T(\xi)\, d\mu(\xi)$$

$$\simeq \int_{\hat{T}_1}^{\oplus} T(\xi)\, d\mu(\xi) \oplus \int_{\hat{T}_2}^{\oplus} T(\xi)\, d\mu(\xi) \oplus \int_{\hat{T}_3}^{\oplus} T(\xi)\, d\mu(\xi) .$$

We have indicated in the previous section that $\hat{T}_1$ and $\hat{T}_2$ may be considered as Borel subsets of R^2 and R^5 respectively and that the binormal operator

$$\int_{\hat{T}_1}^{\oplus} T(\xi)\, d\mu(\xi) \;\oplus\; \int_{\hat{T}_2}^{\oplus} T(\xi)\, d\mu(\xi)$$

may be expressed in the form

$$\int_{\hat{T}_1}^{\oplus} \xi I(\xi)\, d\mu(\xi) \;\oplus\; \int_{\hat{T}_2}^{\oplus} m(\xi) \begin{pmatrix} x(\xi) & y(\xi) \\ 0 & z(\xi) \end{pmatrix} d\mu(\xi)$$

$$\simeq N \oplus \begin{pmatrix} X & Y \\ 0 & Z \end{pmatrix}$$

where N is a normal operator, X, Y and Z are commuting normal operators and Y is a positive operator. Thus we can obtain the corresponding result for trinormal operators by restricting our attention to the third component, the pure J_3 operator

$$R = \int_{\hat{T}_3}^{\oplus} T(\xi)\, d\mu(\xi)\,.$$

Now for each ξ in $\hat{T}_3$, $T(\xi)$ is a type I factor operator of the form $m(\xi)\, S(\xi)$ where $S(\xi)$ is a 3×3 irreducible matrix, which we may suppose is in canonical form. For each positive integer (or $+\infty$) m let

$$\hat{T}_3(m) = \left\{ \xi : \xi \in \hat{T}_3 \text{ and } m(\xi) = m \right\}.$$

The collection of Borel subsets $\hat{T}_3(m)$ of $\hat{T}_3$ is a partition of $\hat{T}_3$ such that

$$R \simeq \sum_m \oplus\, m \int_{\hat{T}_3(m)}^{\oplus} S(\xi)\, d\mu(\xi)$$

where

$$S(\xi) = \begin{pmatrix} a(\xi) & d(\xi) & f(\xi) \\ 0 & b(\xi) & g(\xi) \\ 0 & 0 & c(\xi) \end{pmatrix}.$$

If $\hat{T}_3$ has been parametrized according to corollary 5.4 we may consider each ξ in $\hat{T}_3$ to be a vector in R^{10} and let ξ_i denote the ith component of ξ. Then $a(\xi) = \xi_1 + i\xi_2$, $b(\xi) = \xi_3 + i\xi_4$, $c(\xi) = \xi_5 + i\xi_6$, $d(\xi) = \xi_7$, $g(\xi) = \xi_8$ and $f(\xi) = \xi_9 + i\xi_{10}$. Thus all the maps a, b, c, d, g, f are complex valued Borel functions on $\hat{T}_3$ which are bounded by $\|R\|$. Recall that for each positive integer m, $\mathcal{H}_m$ denotes the Hilbert space of m-tuples of complex numbers and $\mathcal{H}_\infty = \ell_2$. Define the normal operator A on the space

$$\mathcal{H} = \sum_m \oplus\, L^2\left(\hat{T}_2(m), \mu; \mathcal{H}_m\right)$$

by

$$A\left\{\psi_m(\xi)\right\} = \left\{a(\xi)\,\psi_m(\xi)\right\}$$

for $\xi \in \hat{T}_2(m)$, $\psi_m \in L^2\left(\hat{T}_2(m), \mu; \mathcal{H}_m\right)$ and $1 \le m \le +\infty$. Similarly define normal operators B, C, D, G and F as multiplication operators on the space $\mathcal{H}$ corresponding to the bounded Borel functions b, c, d, g, and f respectively. Then

$$\begin{pmatrix} A & D & F \\ 0 & B & G \\ 0 & 0 & C \end{pmatrix} \text{ acts on } \mathcal{H} \oplus \mathcal{H} \oplus \mathcal{H}$$

and is unitarily equivalent to

$$R \simeq \sum_m \oplus \textcircled{m} \int_{\hat{T}_3(m)}^{\oplus} \begin{pmatrix} a(\xi) & d(\xi) & f(\xi) \\ 0 & b(\xi) & g(\xi) \\ 0 & 0 & c(\xi) \end{pmatrix} d\mu(\xi)$$

under the natural identification of $\mathcal{K} \oplus \mathcal{K} \oplus \mathcal{K}$ with

$$\sum_m \oplus \textcircled{m} \left[L^2\left(\hat{T}_2(m), \mu\right) \oplus L^2\left(\hat{T}_2(m), \mu\right) \oplus L^2\left(\hat{T}_2(m), \mu\right)\right].$$

Thus we have proven:

PROPOSITION 5.5. Every trinormal operator T may be written in the form

$$T \simeq N \oplus \begin{pmatrix} X & Y \\ 0 & Z \end{pmatrix} \oplus \begin{pmatrix} A & D & F \\ 0 & B & G \\ 0 & 0 & C \end{pmatrix}$$

where this is a disjoint direct sum, N is a normal operator, X, Y, Z are commuting normal operators, A, B, C, D, G, F are commuting normal operators and Y, D and G are positive operators.

We remark that with propositions 5.3 and 5.5 the techniques of Bunce and Deddens [20] should enable one to compute explicitly the topology of the spectrum $\hat{T}$ of a trinormal operator T, although its description might become somewhat involved.

The unitary equivalence problem for $n \times n$ matrices is solved in various forms [13], [97], [105], [112], [119] and there may well be more general, as well as more natural and appealing ways of describing the Borel space $\hat{T}_n$ (the n-dimensional part of the spectrum $\hat{T}$ of an operator T) than the procedure we have outlined for the case $n = 2$ and $n = 3$. While we know in principle that this method extends to arbitrary n, a specific parametrization is likely to be unreasonably messy, even for the $n = 4$ case. (The reader may have already made such an evaluation of the $n = 3$ case.)

The fact that in general n-normal operators have a decomposition analogous to proposition 5.5 has already been mentioned by Arlen Brown in section 11 of [16].

We conclude with some general observations on the parametrization of the spectrum of an n-normal operator which are implicit in [97] and [105]. If T is an operator then the n-dimensional part $\hat{T}_n$ of the spectrum $\hat{T}$ of T may be parametrized (Borel isomorphically) by a Borel subset of R^{n^2+1}, by means of an appropriate upper-triangular canonical form. In particular if T is an n-normal (J_n) operator, its spectrum $\hat{T}$ may be parametrized (Borel isomorphically) by a Borel subset of real Euclidean space of dimension

$$\sum_{i=1}^{n} \left(i^2 + 1\right) = \frac{2n^3 + 3n^2 + 7n}{6} .$$

We leave the working out of the specific parametrization to confirmed masochists.

5. Quasinormal operators

The examples of the previous two sections dealt only with parametrizing equivalence classes of finite dimensional irreducible operators and then integrating over these Borel spaces to obtain the specific spectral theorem for a particular class of operators. Therefore in this section we'll give an example of the parametrization of a space of unitary equivalence classes of infinite dimensional irreducible operators. Integration over this Borel space will give us the specific form of the spectral theorem (i.e., central decomposition) for a class of operators which we now describe.

Arlen Brown [15] defined an operator T to have property i.n. if T commutes with T^*T. These operators have since been rechristened quasinormal (cf. problem 108 of [78]). (After this section was written, the author received a preprint of [19] which studies the spectrum of a quasinormal operator from a different perspective.)

We first note that the set quasinormal operators has a property in common with the set of normal operators (proposition 2.7).

PROPOSITION 5.6. If $\mathcal{H}$ is a separable Hilbert space, then the set of quasinormal operators in $\mathcal{L}(\mathcal{H})$ is an $\mathfrak{F}_\sigma$ set, relative to the $*$-strong operator topology.

PROOF. Let $r > 0$ and let $\mathcal{L}_r(\mathcal{H})$ denote the ball of $\mathcal{L}(\mathcal{H})$ of radius r, which is $*$-strong closed. If $\{A_\lambda\}$ is a net of quasinormal operators in $\mathcal{L}_r(\mathcal{H})$ converging to A, by proposition 2.2 A is also quasinormal. Thus for each positive integer n, the set of quasinormal operators norm bounded by n is a closed subset of $\mathcal{L}(\mathcal{H})$.

Clearly every operator weakly contained in a quasinormal operator is quasinormal. Thus the spectrum of a quasinormal operator consists of irreducible quasinormal operators. What are they? Every one-dimensional operator is irreducible and quasinormal. Furthermore the unilateral shift S is an irreducible quasinormal operator (cf. corollary to problem 116 and problem 108 of [78]). Thus αS is an irreducible quasinormal operator for every $\alpha \in \mathbb{C}$, $\alpha \neq 0$. Surprisingly the list (up to unitary equivalence) of quasinormal irreducible operators ends there. Indeed it follows from theorem 1 of [15] that an irreducible quasinormal operator is either normal (and hence one dimensional) or what Brown calls a dilated shift operator defined by a positive operator. In the latter case the irreducibility implies that the dilated shift must have the simple form αS where α is a positive number. What happened to the operators of the form αS where α is complex but not positive?

LEMMA 5.7. If S denotes the unilateral shift and α and β are complex numbers, then $\alpha S \simeq \beta S$ if and only if $|\alpha| = |\beta|$.

PROOF. The "only if" part is obvious as unitarily equivalent operators must have the same norm. The other direction follows from theorem 1, page 15 of [59] since that result can be used to show that if $|\alpha| = 1$ then $\alpha S \simeq S$.

COROLLARY 5.8. Quasinormal operators are smooth.

PROOF. We have all the irreducible operators that can possibly be weakly contained in a quasinormal operator exhibited before us and they certainly satisfy condition 4 of proposition 1.53.

If T is quasinormal, the one-dimensional part $\hat{T}_1$ of $\hat{T}$ (i.e., the normal spectrum of T) is of course parametrized by a compact subset of $\mathbb{C}$ (cf. remarks following corollary 2.48). All the other finite dimensional parts of $\hat{T}$ are empty.

Let $\mathcal{K}_\infty = \ell_2$ and let S denote the concrete unilateral shift on ℓ_2. Then $\{\alpha S : \alpha > 0\}$ is a Borel subset of $\mathcal{L}(\mathcal{K}_\infty)$ and hence a standard Borel space. (This follows for example from theorem 3.2 of [96] since scalar multiplication is $*$-strong continuous and thus the map

$R^+ \times \{S\} \to \{\alpha S : \alpha \in R^+\}$ is a one-to-one Borel map into the standard Borel space $\mathcal{L}(\mathcal{H}_\infty)$.) Thus the quotient map (with respect to unitary equivalence) is a one-to-one Borel map of the standard Borel space $T^c \cap \{\alpha S : \alpha > 0\}$ onto the standard Borel space $\hat{T}_\infty \subset \hat{T}$ ($\hat{T}$ is a standard Borel space by corollary 5.8 and theorem 2.32 of this memoir). Again by theorem 3.2 of [96] this map is a Borel isomorphism onto $\hat{T}_\infty$. We have thus proved:

PROPOSITION 5.9. (Cf. theorem 1 and 2 of [19]) If T is a quasinormal operator, then $\hat{T}_1$ is the normal spectrum of T, $\hat{T}_n$ is empty for $1 < n < +\infty$ and $\hat{T}_\infty$ is parametrized (Borel isomorphically) by a bounded Borel subset of the positive real line. Specifically if $\xi \in \hat{T}_\infty$ and r is the corresponding parameter in R^+, then the unitary equivalence class ξ contains the operator rS, where S is the unilateral shift.

Having described the spectrum of a quasinormal operator, we next examine what the spectral theorem for a quasinormal operator T looks like. Suppose the central decomposition of T is given by

$$T \simeq \int_{\hat{T}}^{\oplus} T(\xi)\, d\mu(\xi)$$

$$(1) \qquad \simeq \int_{\hat{T}_1}^{\oplus} \xi I(\xi)\, d\mu(\xi) \oplus \int_{\hat{T}_\infty}^{\oplus} T(\xi)\, d\mu(\xi) .$$

Here $\hat{T}_1$ is the normal spectrum of T and

$$\int_{\hat{T}_1}^{\oplus} \xi I(\xi)\, d\mu(\xi)$$

is just the normal kernel of T in its spectral decomposition form (cf. section 2 of this chapter). Thus we restrict our attention to the second direct summand of (1) and henceforth consider $\hat{T}_\infty$ as parametrized by a bounded Borel set of positive numbers. Thus the second direct summand of (1) is

$$Q = \int_{\hat{T}_\infty}^{\oplus} T(r)\, d\mu(r)$$

which could be called the reduced quasinormal operator or a pure quasinormal operator. Each $T(r)$ is a type I factor operator weakly contained in T and hence is of the form $T(r) \simeq \textcircled{m(r)}(rS)$ for some positive integer (or $+\infty$) $m(r)$, where S as usual denotes the unilateral shift. Let $\hat{T}_\infty(m) = \left\{ r : r \in \hat{T}_\infty \text{ and } m(r) = m \right\}$ for $1 \le m \le +\infty$. By lemma 4.3 $\left\{ \hat{T}_\infty(m) \right\}$ is a countable Borel partition of $\hat{T}_\infty$. Hence the spectral decomposition of Q may be put into the form

$$Q \simeq \sum_m \oplus \textcircled{m} \int_{\hat{T}_\infty(m)}^{\oplus} r\, S\, d\mu(r) .$$

Except for the presence of the shift S in the integrand, this looks very much like the spectral theorem for a positive operator. Indeed in Arlen Brown's study [15] Q is unitarily isomorphic to a dilated shift associated with a positive operator P_o. To make that connection specific we mention that the appropriate positive operator P_o is precisely the operator

$$P_o = \sum_m \oplus \textcircled{m} P_m$$

where each P_m is a multiplicity free positive operator defined on $L^2\left(\hat{T}_\infty(m), \mu\right)$ by

$$(P_m f)(r) = rf(r) \quad \text{for all } f \text{ in } L^2\left(\hat{T}_\infty(m), \mu\right), \quad r \in \hat{T}_\infty(m).$$

This method may be used to obtain the specific form of the spectral theorem for operators A which commute with AA^*. This is the class of all adjoints of quasinormal operators and hence, for example, we may apply proposition 2.42 to obtain the adjoint image of proposition 5.9.

PROPOSITION 5.10. If T commutes with TT^* then $\hat{T}_1$ is the normal spectrum of T, $\hat{T}_n$ is empty for $1 < n < +\infty$, and $\hat{T}_\infty$ is parametrized (Borel isomorphically) by a bounded Borel subset of the positive real line. Specifically if $\xi \in \hat{T}_\infty$ and r is the corresponding parameter in $\mathbb{R}^+$, then the unitary equivalence class ξ contains the operator rS^*, where S^* is the backwards shift.

6. Direct sum decompositions

By the definition of the central decomposition of an operator T, whenever B is a Borel subset of its quasi-spectrum $\tilde{T}$ we may write T as a direct sum

$$T \simeq \int_B^\oplus T(\xi)\, d\mu(\xi) \oplus \int_{\tilde{T}-B}^\oplus T(\xi)\, d\mu(\xi)$$

where the two direct summands are disjoint operators. Often we know that certain basic classes of operators form a Borel subset (for example type I, type II, type III or n-dimensional operators) of $\tilde{T}$, which will then give an interesting direct sum decomposition of T into disjoint components. Propositions 1.19, 1.23, 1.29 and 5.5 are all examples of this phenomenon.

Proposition 5.5 asserts (among other things) that every trinormal operator T is the direct sum of a normal operator (the normal kernel of T), a pure J_2 (or homogeneous binormal) operator and a pure J_3 (or homogeneous trinormal) operator. Since the n-dimensional parts $\hat{T}_n$ of $\hat{T}$, where n is a positive integer, together with the remaining set $\tilde{T}_\infty$ of strictly infinite dimensional elements of $\tilde{T}$, form a Borel partition of $\tilde{T}$, we obtain the following disjoint direct sum decomposition of any operator.

PROPOSITION 5.11. Every operator T has a unique disjoint direct sum decomposition of the form

$$T \simeq \sum_{1 \le n \le +\infty} \oplus\, T(n)$$

where $T(1)$ is the normal kernel of T, $T(n)$, $1 < n < +\infty$ is a pure J_n (or homogeneous n-normal) operator and $T(+\infty)$ has no n-normal suboperators for any n, $1 \le n < +\infty$. (Of course not all terms in the direct summand need occur.)

We have proved (cf. proposition 5.6) that the set of quasinormal elements of $\tilde{T}$ is a Borel subset of $\tilde{T}$. Hence we obtain the following decomposition of any operator.

PROPOSITION 5.12. Every operator T has a unique disjoint direct sum decomposition of the form

$$T = T_q \oplus T_o$$

where T_q is a quasinormal operator and T_o has no quasinormal suboperators.

We call T_q the quasinormal kernel of T and call T_o a completely nonquasinormal operator. Further one can apply proposition 5.12 to the term

$T(+\infty)$ in proposition 5.11 to obtain a refinement of that decomposition. The quasinormal kernel of $T(+\infty)$ would then have no normal suboperators and hence might be called a pure quasinormal operator. Thus whenever some basic (Borel) portion of the quasi-spectrum of an operator is sufficiently understood, one can use this technique to concentrate on the area of ignorance — a "reduced operator."

We mention two more classical examples which are subsumed by this point of view and then the reader can adapt this method to his (or her) own needs. The portion of the one-dimensional part $\hat{T}_1$ of an operator T that lies on the real line is a Borel subset of $\hat{T}_1$ and hence of $\tilde{T}$. Thus every operator has a unique disjoint direct sum decomposition into a Hermitian operator and an operator which has no Hermitian suboperators (this is theorem 1.2 of [14]). The first operator is called the Hermitian kernel of T and the second is called a completely non-Hermitian operator. Similarly the intersection of $\hat{T}_1$ with the unit circle is a Borel subset of $\hat{T}_1$ and hence of $\tilde{T}$. Thus T has a unique disjoint direct sum decomposition into a unitary operator and an operator which has no unitary suboperators. Again we call the first operator the unitary kernel of T and the second a completely non-unitary operator. This is just theorem 3.2 of [60] except that there it is stated and proven only for contraction operators.

7. Some research proposals

In this memoir we have not tried to chart the realm of operators, but rather to set up one possible mapping procedure for bringing some organizational sense to the enormous task of exploring and describing the vast and varied operator terrain. Here we are using the term "mapping" not in the sense of the mathematician but in the sense of the geographer. Cartographers often have various methods for describing the same area. A map perfectly suited to the needs of a geologist may be all but useless to the backpacker. We of course hope that our suggested procedure, involving quasi-spectra, central measure classes and multiplicity functions, will prove suitable as a framework for many operator theoretic studies. As a cartographic method, this memoir suggests far more questions than it answers, and as such has more the character of a research proposal than a treatise. We conclude by distilling from the many questions that come to mind, eight specific areas of potential research activity.

a. Colonizing smooth territories. Here the job is to describe, as specifically as possible, various Borel spaces of unitary equivalence classes of irreducible operators (and in some cases Borel spaces of quasi-equivalence classes of factor operators). Even the situation for $n \times n$ matrices is not satisfactory (cf. sections 3 and 4 of this chapter). If a matrix theory expert could devise a new method (perhaps not even involving an upper triangular form) which would give a natural Borel cross-section of the unitary equivalence classes of irreducible $n \times n$ matrices and have a reasonable and specific description in the general $n \times n$ case he (or she [55]) would make n-normal operators (almost) as easy to work with as normal operators.

A somewhat broader area (which includes matrices) which would be interesting to characterize and possibly parametrize is the Borel space of quasi-equivalence classes of cofinite factor operators. (Cf. definition 2.45, page 78.) Alain Guichardet has shown that this is a smooth (i.e., standard) Borel space (cf. proposition 2.46 of this memoir and theorem 1, page 21 of [74]). The type I cofinite operators are (up to quasi-equivalence) the finite dimensional irreducible operators so that the real unknowns here are the operators which generate type II_1 factor von Neumann algebras. The solution of the unitary equivalence problem for $n \times n$ matrices [97], [105] and [119], has enabled Carl Pearcy to describe a complete set of unitary invariants for cofinite type I operators [106].

Considerable work has been done on the unitary equivalence problem for various other classes of operators: partial isometries [80], Hilbert-Schmidt operators [34], [35], compact operators [34], [2], [6] for example. Such results might be used, for instance, to describe the Borel space of (unitary equivalence classes of) irreducible compact operators. Integrating over such spaces should then enable one to study and classify, up to unitary equivalence a rather large class of operators, namely those operators whose spectra consist entirely of compact operators. It is not difficult to see that this is the important class of operators which generate (without identity) a C^*-algebra which is CCR in the Kaplansky terminology [88] or "liminaire" in the Dixmier terminology (definition 4.2.1, page 86 of [36]). Since we are unsatisfied with the C^*-algebra terminology and do not wish to be guilty of aiding its perpetuation in a new context, we here christen these operators "K-normal." The rationale for this terminology is follows. We define an operator T to be n-normal if all the irreducible operators weakly contained in T are finite dimensional, of dimension at most n. Thus an operator is K-normal if all the irreducible operators weakly contained in T are compact. The problem immediately arises, of characterizing which operators are in this very special class. Recall (theorem 1, page 582 of [69]) that the work of James Glimm gives a characterization of smooth operators in terms of the topology of the spectrum. Specifically an operator T is smooth if and only if $\hat{T}$ has T_0 - topology. It also follows from the work of James Glimm (theorem 4, page 599 of [69] that the K-normal operators may be characterized in terms of the topology of the spectrum. Indeed James Glimm has characterized CCR algebras as precisely those whose spectrum has T_1 - topology. Here we run into a minor technical difficulty as an operator T is K-normal if and only if the C^* - algebra $\mathcal{A}$ generated by T, <u>without the identity</u> is CCR, while the spectrum $\hat{T}$ of T corresponds to the spectrum of the C^*-algebra $C^*(T)$ generated by T <u>and</u> the identity operator. However the spectrum of $\mathcal{A}$ and of $C^*(T)$ are closely related. (Cf. 3.2.4 and 3.9.5 of [36].) If $\mathcal{A}$ contains the identity, then all the irreducible operators weakly contained in T must be finite dimensional and $\hat{T}$ has T_1 - topology. If $\mathcal{A}$ does not contain the identity then T weakly contains the one-dimensional zero operator 0. Thus $(\hat{T} - \{0\})$ has T_1 - topology (in the relative topology) and the closure of each point s in $\hat{T}$ is $\{s,0\}$. Conversely if either $\hat{T}$ has T_1 - topology or $0 \in \hat{T}$ and $\hat{T} - \{0\}$ has T_1 - topology, then T is K-normal.

The task of characterizing these two special classes of operators (K-normal operators and smooth operators) is of great importance. These two problems correspond to questions which have already led to extensive and deep results in topological group theory, namely: which groups are CCR and which groups are type I ?

b. Exploring rough territories. I am sure there are those who consider the first mentioned research area an advanced form of library science and who would prefer to explore the mysterious world of nonsmooth operators. Here even rare sightings of strange creatures are of interest. One such sighting is the Topping operator mentioned in proposition 1.53, which has been concretely represented by John Bunce and James Deddens. (Cf. remark following corollary 8 of [23].) This Topping operator is my candidate for a transitive operator, i.e., a bounded operator admitting no proper closed invariant subspaces.

Nonsmooth operators may not be all that inaccessible. The meagerness of our knowledge and the lack of many concrete examples may simply be due to the fact that we have not been looking for them. As we mentioned earlier it follows from the work of Alain Guichardet [74] that the space of quasi-equivalence classes of cofinite type II factor operators (all nonsmooth operators) is a smooth (i.e., standard) Borel space. Further John Bunce and James Deddens have recently discovered nonsmooth operators hiding among the weighted shifts [21], [23] (cf. also [101]). Further Warren Wogen [131], [132] has shown there exist hyponormal operators and nilpotent operators which are not smooth. Also there exist partial isometries which are not smooth [107], [116]. This remains a largely unexplored area of research and we suspect many of those nonsmooth operators have some surprises in store for us.

An area of recent mathematical activity has been the study of type III factor von Neumann algebras. All of these are singly generated. Hence much of this work should stimulate an interest in type III factor operators. For example Alain Connes [29] has introduced a new algebraic invariant such that the von Neumann classification of factors can be further refined. Thus Connes speaks of factors of type III_λ for $0 \leq \lambda \leq 1$. Can we translate this new classification into operator theoretic terms so that we can give a natural definition of type III_λ operators, just as we did for the ordinary von Neumann classification of factors? Both Masamichi Takesaki and Alain Connes [29] have developed structure theories for type III factor von Neumann algebras. Can one use these theories to get concrete examples of operators of each of the types III_λ $0 \le \lambda \le 1$? While we know that the classification of all factor operators is not smooth (corollary 2.33) is it conceivable that the classification of type III factor operators is smooth? This example exhibits a general research thrust. Whenever new and exciting results are obtained in the classification of factor von Neumann algebras, they could well generate new questions and insights in the study of single operators.

c. Unbounded operators and nonseparable spaces. For simplicity and clarity we have restricted our attention in this memoir to bounded operators on separable spaces, even when that was not strictly necessary. For example, as Harry Gonshor has pointed out at the end of [71] the spectral theorem for n-normal operators can be worked out for unbounded operators as well. Thomas Chow [25] has also worked out some of the direct integral theory for unbounded (closed) operators. Certainly the whole theory will be more utilitarian if large parts of it are extended to appropriate classes of unbounded operators.

Similarly there are versions of the ordinary spectral theorem for normal operators which are valid for nonseparable spaces (cf. page 60 of [78]). It would be useful to have similar versions of the general spectral theorem for nonseparable spaces. After all it is occasionally interesting to study operator theory on nonseparable spaces [44]. Even when the space is nonseparable, we are fortunate that the C^*-algebras to be considered are singly generated and hence separable. By contrast the C^*-algebra of all operators on a separable infinite dimensional Hilbert space, is not separable.

d. Borel structure. This is a highly technical aspect of the subject where much remains to be done. As we have indicated in section 6 of this chapter, whenever a major class of operators (compact? hyponormal? subnormal?) can be proven to be a Borel set we obtain a corresponding disjoint direct sum decomposition of arbitrary operators. To complicate matters further the spectrum $\hat{T}$ of an operator T in general has three Borel structures, the Mackey Borel structure, the Davies Borel structure and the topological Borel structure. Any and all technical information as to which sets, and which operations, are Borel relative to which Borel structures, is likely to be very useful. For example, any specific application of the characterization (theorem 3.18) of the central measures will entail a good working knowledge of both the Mackey and the Davies Borel structures in the quasi-spectrum $\tilde{T}$ of an operator.

e. Topological structure. In this memoir, with our primary interest in direct integral theory, we have slighted somewhat an area of research that is potentially most intriguing, namely the study of the (in general nonHausdorff) topology of the spectrum of an operator and its relationship with the properties of the operator. John Bunce and James Deddens have recently computed the topology of the spectrum of a binormal operator [20] (cf. also [19]). C^* - algebra theoreticians are also eager to see progress in this field as they too are short on examples where the complete spectrum of a C^* - algebra has been computed as a topological space.

As we indicated in the discussion of the first research area (colonizing smooth territories) various separation properties of the topology of the spectrum correspond to basic classes of operators (K-normal and smooth). It is therefore natural to ask the question as to which operators have Hausdorff spectra. Indeed John Bunce and James Deddens [24] have made substantial progress towards characterizing this special class of operators. (See also corollary 2, page 388 of [57] or theorem 4.2 of [88].)

One way to get greater insight into the topology of the spectrum is to consider various alternate descriptions of the topology (cf. for example [54]). Our description of the topology is in terms of weak containment (definition 2.52). John Bunce and James Deddens [22] have defined a spatial notion which is closely related to the notion of weak containment. If S and T are operators on Hilbert spaces $\mathcal{H}$ and $\mathcal{H}'$ respectively, then S is a <u>subspace approximant</u> to T if, for every $\epsilon > 0$, there is an operator R_ϵ on a space $\mathcal{H}_\epsilon$ and a unitary transformation U_ϵ of $\mathcal{H}$ onto $\mathcal{H}' \oplus \mathcal{H}_\epsilon$ such that

$$\|T - U_\epsilon^*(S \oplus R_\epsilon)U_\epsilon\| < \epsilon .$$

($\mathcal{H}_\epsilon$ can be trivial, i.e., zero dimensional.) This is equivalent to saying that T is a norm limit of operators having suboperators unitarily equivalent to S. Bunce and Deddens [22] prove that if S is a subspace approximant of T then S is weakly contained in T. They note that the converse is clearly false in general, but conjecture that it is true when S is irreducible. They verify the conjecture under certain conditions, such as when S is finite dimensional or when T is quasi-normal. Donald Hadwin has shown the conjecture is true when the irreducible operator S is compact. (Cf. Theorem 3.3 of [76].) The relationship of "subspace approximant" and "weak containment" needs to be further explored. While the notion of weak containment is a very algebraic one, it would be most useful to obtain a more geometric characterization of the concept. (We mention that the work of the author referred to in [22], [24], and [76] is an earlier version of this memoir, under a different title.)

f. <u>Adaptation of representation theory to the single operator context.</u> In this memoir we have modified and specialized many notions and theorems from $*$-representation theory and C^*-algebras and reformulated these results in the operator context. Following this general pattern there are undoubtedly many more representation theoretic concepts which would take an interesting form if reinterpreted in terms of single operators. The treatise by Jacques Dixmier [36], so heavily used in this memoir, has much more material which would be relevant to the study of individual operators. In addition there are many aspects of representation theory (for example induced representations) which are not covered in the treatise. Furthermore the field of representation theory has grown enormously since that treatise came out in 1964.

There is a simple device for translating representation theoretic results directly into the operator context. Indeed those familiar with representation theory can use this procedure to obtain major portions of this memoir. We define a universal separable C^*-algebra, $C^*(z,z^*)$ as follows. Let $F(z,w)$ be the free complex algebra on two non-commuting variables z, w. The exchange $(z,w) \to (w,z)$ gives rise to a unique involution $*$ of $F(z,w)$ such that $w = z^*$. We define a norm on $F(z,w)$ by

$$\|p(z,z^*)\| = \sup_{\|T\| \le 1} \|p(T,T^*)\|$$

where p is a complex polynomial contained in $F(z,w)$ and the supremum is taken over all contraction operators on a separable Hilbert space. $C^*(z,z^*)$ is defined to be the completion of $F(z,z^*)$ with respect to this norm. Then each contraction operator T determines a $*$-representation π_T of $F(z,z^*)$ by $\pi_T(p) = p(T,T^*)$. Clearly π_T has a unique extension to $C^*(z,z^*)$. In this way one gets a bijective correspondence between the set of contraction operators on a separable Hilbert space $\mathcal{H}$ and the collection of $*$-representations of $C^*(z,z^*)$ acting on $\mathcal{H}$. This correspondence then respects all the important notions such as unitary equivalence, quasi-equivalence, weak equivalence, type classification, etc.

A major theory in group representations which has not been adapted to the context of operators is the theory of induced representations, particularly as developed by George Mackey [94], [95]. This theory has played a fundamental role in computing the duals (= spectra) of topological groups. Any general procedure for manufacturing irreducible operators would be useful, particularly for the area of research first mentioned in this list. While the suggestion is admittedly vague in the extreme there is nevertheless some basis for optimism as Mackey's method has been generalized to various non-group situations. There is the recent work of Marc Rieffel extending the method to $*$-representations of C^*-algebras [113], [114]. Further Michael Fell has extended the theory to the general setting of "Banach $*$-algebraic bundles" [58].

While Mackey's theory of induced representations is one of the obvious examples to try, there are undoubtedly large parts of C^*-algebra theory which could be modified to give information about operators. This memoir has concentrated only on the most basic and immediately adaptable portions of representation theory.

g. Spectral theory. The ordinary (numerical) spectrum of an operator has been studied up-side down and side-ways; it has been partitioned into various interesting subsets, it has been related to the numerical range, it has been perturbated, etc. How many of these well developed techniques can now be extended to the more complicated spectrum introduced in this memoir? Certainly the few results of section 4 of chapter 2 hardly scratch the surface of this area of investigation. For example we can say that an operator S is essentially contained in an operator T if there exists a $*$-representation φ of the quotient C^*-algebra $C^*(T)/K \cap C^*(T)$ (where K is the ideal of compact operators on the space $\mathcal{H}(T)$ on which T acts) such that $\varphi(\varphi_o(T)) = S$, where φ_o denotes the quotient map of $C^*(T)$ onto $C^*(T)/K \cap C^*(T)$. We could then define $\hat{T}_e$, the essential spectrum of T, to be the set of unitary equivalence classes of irreducible operators essentially contained in T. Then $\hat{T}_e$ is a non-empty compact (but not Hausdorff) subset of the spectrum $\hat{T}$, which reduces to the ordinary (i.e., scalar) essentially spectrum when T is normal.

Recently there have been other attempts to generalize the notion of spectrum to a "spectrum" which contains operators rather than merely scalars. For example Carl Pearcy and Norberto Salinas ([108], [109] and [110]) define the $n \times n$ reducing spectrum of T, denoted $R^n(T)$, to be the set of all $n \times n$ matrices weakly contained in T. Similarly the $n \times n$ reducing essential spectrum, denoted $R_e^n(T)$, is the set of all $n \times n$ matrices essentially contained in T. For n-normal operators these notions are related to the spectra introduced in this memoir as follows. Our spectrum $\hat{T}$ of an n-normal operator T is just the quotient (with respect to unitary equivalence) of the set of irreducible elements of $\bigcup_{m=1}^{n} R^m(T)$. Similarly the essential spectrum $\hat{T}_e$ defined above is the quotient (with respect to unitary equivalence) of the set of irreducible elements of $\bigcup_{m=1}^{n} R_e^m(T)$. Thus if S and T are n-normal operators then $\hat{S} \subset \hat{T}$ if and only if $R^n(S) \subset R^n(T)$ and $\hat{S}_e \subset \hat{T}_e$ if and only if $R_e^n(S) \subset R_e^n(T)$. Thus various results of Carl Pearcy and Norberto Salinas on compact perturbations of n-normal operators (theorems 8, 9 and 10 of [108]) may be stated equally as well in terms of the spectra $\hat{T}$ and $\hat{T}_e$.

Donald Hadwin, in section 3 of his thesis [76], has also introduced generalized spectral notions. He defines the <u>reducing operator spectrum</u> (denoted $\Sigma(T)$) of an operator T to be the set of operators which are subspace approximants of T, in the sense of John Bunce and James Deddens [22]. (The notion of subspace approximant is defined and discussed above under research proposal 5, topological structure.) Further the <u>essential reducing operator spectrum</u> of T is defined by Hadwin to consist of those operators S for which $\textcircled{\infty} S$ is in the reducing operator spectrum of T. (Some specific choices of Hilbert spaces are needed here in order to make the above spectra well defined sets. In our approach we avoid such an arbitrary choice by using sets of unitary equivalence classes.) The quotient $H(T)$ (with respect to unitary equivalence) of the irreducible operators in $\Sigma(T)$ is then a subset of $\hat{T}$, and the assertion that $\hat{T} = H(T)$ is equivalent to the Bunce-Deddens conjecture that "weak containment" and "subspace approximant" are equivalent for irreducible operators. Since Hadwin has verified the conjecture for compact irreducible operators (Theorem 3.3 of [76]) and since any compact operator is a direct sum of irreducible compact operators, it is clear that, for K-normal operators, $\hat{T} = H(T)$ and $\hat{S} \subset \hat{T}$ if and only if $\Sigma(S) \subset \Sigma(T)$. It therefore seems quite likely that the techniques and results of Donald Hadwin could be used to obtain an extension of the Halmos-Berg-Pearcy-Salinas results on compact perturbations of n-normal operators (for example theorem 9 of [108]) to K-normal operators.

As another fascinating example of the extension of scalar spectral concepts to spectra consisting of irreducible operators, we mention that Donald Hadwin in his thesis [76] has introduced the appropriate operator analogues of eigenvalue, multiplicity of an eigenvalue, eigenspace, and isolated eigenvalue.

We conclude this section with one more example of how spectral theory for normal operators can now be extended to our generalized theory for non-normal operators. John Conway and Pei Yuan Wu [30] have noted, in the introduction of their paper, the folk theorem that states if T_1 and T_2 are two normal operators then the von Neumann algebra generated by $T_1 \oplus T_2$ splits into the direct sum of the von Neumann algebras generated by T_1 and T_2 if and only if T_1 and T_2 have mutually singular spectral measures. In terms of our generalized spectral theory this fact remains true for non-normal operators as well, i.e., the von Neumann algebra generated by $T_1 \oplus T_2$ splits into the direct sum of the von Neumann algebras generated by T_1 and T_2 if and only if T_1 and T_2 have mutually singular central measures. Indeed (cf. section 2.2 of chapter 1 of [37]) such a splitting occurs if and only if the projection E of the space $\mathcal{H}(T_1 \oplus T_2)$ of $T_1 \oplus T_2$ onto the space $\mathcal{H}(T_1)$ of T_1, is in the center of $\mathcal{A}(T_1 \oplus T_2)$. By part 3 of proposition 1.35 this occurs if and only if T_1 and T_2 are disjoint. The result now follows from corollary 3.12.

h. Is the bridge two-way? The emphasis of this memoir has been the adaptation of the techniques and concepts of the representation theory of C^* - algebras to the operator theoretic context. Once this connection is better understood, operator techniques may well contribute significantly to the theory of C^* - algebras. For example any progress in extending operator theoretic spectral theory to the more general spectra considered in this memoir, will likely apply as well to spectra of C^* - algebras. Thus the research that has been done on which operators have nonempty normal spectra relates to the whole question of which C^* - algebras admit characters. One could define the essential spectrum of a general C^* - algebra to be the subset of the ordinary spectrum consisting of those (unitary equivalence classes of irreducible) representations which factor through the Calkin algebra. Can any of the interesting theory surrounding the concept of essential spectrum for operators be extended to this C^*-algebra context? The recent thrust of research in the direction of joint spectra of sets of operators brings us closer to dealing with arbitrary C^*-algebras rather than just singly generated ones [18], [92]. On the other hand operator theory corresponds to a special class of C^*-algebras, namely the singly generated ones. It would be useful to characterize that class and to study the structure of the algebras in the class in terms of their generating element. (Catherine Olsen and William Zame [103] have exhibited various kinds of C^*-algebras which they show are singly generated.) Perhaps the area of greatest potential utility of operator theory to C^*-algebra theory is the construction and analysis of specific examples. As a case in point, John Bunce and James Deddens [20] have recently computed the spectra (with topology) of certain C^*-algebras by essentially operator theoretic techniques (cf. also [19]).

REFERENCES

1. Charles Akemann, The dual space of an operator algebra, Trans. Amer. Math. Soc. 126 (1967), 286-302. MR34 # 6549

2. William Arveson, Unitary invariants for compact operators, Bull. Amer. Math. Soc. 76 (1970), 88-91. MR40 # 4803

3. ______, An invitation to C^* - algebras, to appear, Graduate texts in Mathematics, Springer-Verlag, New York.

4. ______, Operator algebras and invariant subspaces, Annals of Math. (2) 100 (1974), 433-532.

5. ______, Subalgebras of C^* - algebras, Acta Mathematica, 123 (1969), 141-224. MR40 # 6274

6. ______, Subalgebras of C^* - algebras II, Acta Mathematica 128 (1972), 271-308.

7. Horst Behncke, Generators of W^* - algebras, Tôhoku Math. J. 22 (1970), 541-546.

8. ______, Generators of W^* - algebras II, Tôhoku Math. J. 24 (1972), 371-381. MR48 # 4747

9. ______, Generators of W^* - algebras III, Tôhoku Math. J. 24 (1972), 383-388. MR48 # 4747

10. ______, Structure of certain nonnormal operators II, Indiana Univ. Math. J. 22 (1972), 301-308. MR47 # 9322

11. Sterling K. Berberian, Approximate proper vectors, Proc. Amer. Math. Soc. 13 (1962), 111-114. MR24 # 3516

12. ______, A note on hyponormal operators, Pacific J. Math. 12 (1962), 1171-1175. MR26 # 6771

13. Joel Brenner, The problem of unitary equivalence, Acta Math. 86 (1951), 297-308. MR13, 717

14. Mikhail Samoǐlovich Brodskii, Triangular and Jordan representations of linear operators [Russian], Izdat. "Nauka," Moscow 1969. MR41 # 4283. See also Translations of Math. Monographs, Vol. 32, Amer. Math. Soc., Providence R.I., 1971.

15. Arlen Brown, On a class of operators, Proc. Amer. Math. Soc. 4 (1953), 723-728. MR15, 538

16. ______, The unitary equivalence of binormal operators, Amer. J. Math. 76 (1954), 414-434. MR15, 967

17. John Bunce, Characters on singly generated C^* - algebras, Proc. Amer. Math. Soc. 25 (1970), 297-303. MR41 # 4258

18. ______, The joint spectrum of commuting nonnormal operators, Proc. Amer. Math. Soc. 29 (1971), 499-505. MR44 # 832

19. ______, Irreducible representations of the C^* - algebra generated by a quasi-normal operator, Trans. Amer. Math. Soc. 183 (1973), 487-494. MR48 # 4805

20. John Bunce and James Deddens, Irreducible representations of the C^*-algebra generated by an n-normal operator, Trans. Amer. Math. Soc. 171 (1972), 301-307. MR46 # 6051

21. ______, C^*-algebras generated by weighted shifts, Indiana Univ. Math. J. 23 (1973), 257-271. MR47 # 5858

22. ______, Subspace approximants and GCR operators, Indiana Univ. Math. J. 24 (1974/75), 341-349.

23. ______, A family of simple C^*-algebras related to weighted shift operators, J. Functional Analysis 19 (1975), 13-24.

24. ______, C^*-algebras with Hausdorff spectrum, to appear, Trans. Amer. Math. Soc.

25. Thomas R. Chow, A spectral theory for direct integrals of operators, Math. Ann. 188 (1970), 285-303. MR42 # 3598

26. ______, The spectral radius of a direct integral of operators, Proc. Amer. Math. Soc. 26 (1970), 593-597. MR42 # 3598

27. Louis Coburn, The C^*-algebra generated by an isometry, Bull. Amer. Math. Soc. 73 (1967), 722-726. MR35 # 4760

28. ______, The C^*-algebra generated by an isometry II, Trans. Amer. Math. Soc. 137 (1969), 211-217. MR38 # 5015

29. Alain Connes, Une classification des facteurs de type III, Ann. Sci. École Norm. Sup. (4) 6 (1973), 133-252. MR49 # 5865

30. John B. Conway and Pei Yuan Wu, The splitting of $\sigma(T_1 \oplus T_2)$ and related questions, preprint.

31. E. Brian Davies, On the Borel structure of C^*-algebras, Commun. Math. Phys. 8 (1968), 147-163. MR37 # 6764

32. ______, Decomposition of traces on separable C^*-algebras, Quart. J. Math. Oxford (2) 20 (1969), 97-111. MR39 # 1984

33. ______, The structure of Σ^*-algebras, Quart. J. Math. Oxford (2) 20 (1969), 351-366. MR41 # 835

34. Don Deckard, Complete sets of unitary invariants for compact and trace-class operators, Acta Sci. Math. (Szezed) 28 (1967), 9-20. MR36 # 732

35. Don Deckard and Carl Pearcy, On unitary equivalence of Hilbert-Schmidt operators, Proc. Amer. Math. Soc. 16 (1965), 671-675. MR31 # 3866

36. Jacques Dixmier, Les C^*-algèbres et leurs représentations, Gauthier-Villars, Paris, 1964. MR30 # 1404

37. ______, Les algèbres d'opérateurs dans l'espace Hilbertien (Algèbres de von Neumann), Gauthier-Villars, Paris, 1969. MR20 # 1234

38. ______, Dual et quasi-dual d'une algèbre de Banach involutive, Trans. Amer. Math. Soc. 104 (1962), 278-283. MR25 # 3384

39. ______, Utilization des facteurs hyperfinis dans la théorie des C^*-algèbres, Comptes Rendus Acad. Sc. Paris 258 (1964), 4184-4187. MR28 # 4382

40. ______, Sur les C^*-algèbres, Bull. Soc. Math. Fr. 88 (1960), 95-112. MR22 # 12408

41. ______, Sur les structures Borélinennes du spectre d'une C^* - algèbre, Publications Math., Inst. des hautes études scientifiques, No. 6 (1960), 297-303. MR23 # A2065

42. Ronald Douglas, Banach algebra techniques in operator theory, Academic Press, New York, 1972.

43. Ronald Douglas and Carl Pearcy, Von Neumann algebras with a single generator, Michigan Math. J. 16 (1969), 21-26. MR39 # 6089

44. Gerald Edgar, John Ernest and Sa Ge Lee, Weighing operator spectra, Indiana Univ. Math. J. 21 (1971), 61-80.

45. Edward Effros, A decomposition theory for representations of C^* - algebras, Trans. Amer. Math. Soc. 107 (1963), 83-106. MR26 # 4202

46. ______, The Borel space of von Neumann algebras on a separable Hilbert space, Pacific J. Math. 15 (1965), 1153-1164. MR32 # 2923

47. ______, Global structure in von Neumann algebras, Trans. Amer. Math. Soc. 121 (1966), 434-454. MR33 # 585

48. ______, The canonical measures for a separable C^* - algebra, Amer. J. Math. 92 (1970), 56-60. MR41 # 4259

49. Masatoshi Enomoto, Masatoshi Fujii and Kazuhiro Tamaki, On normal approximate spectrum, Proc. Japan Acad. 48 (1972), 211-215. MR47 # 851

50. John Ernest, A decomposition theory for unitary representations of locally compact groups, Trans. Amer. Math. Soc. 104 (1962), 252-277. MR25 # 3383

51. ______, A new group algebra for locally compact groups, Amer. J. Math. 86 (1964), 467-492. MR29 # 4838

52. ______, A new group algebra for locally compact groups II, Canad. J. Math. 17 (1965), 604-615. MR32 # 159

53. ______, The representation lattice of a locally compact group, Illinois J. Math. 10 (1966), 127-135. MR32 # 1288

54. ______, On the topology of the spectrum of a C^* - algebra, Math. Ann. 216 (1975), 149-153.

55. ______, Mathematics and Sex, to appear, Amer. Math. Monthly.

56. J. Michael Fell, C^* - algebras with smooth dual, Illinois J. Math. 4 (1960), 221-230. MR23 # A2064

57. ______, The dual spaces of C^* - algebras, Trans. Amer. Math. Soc. 94 (1960), 365-403. MR26 # 4201

58. ______, An extension of Mackey's Method to Banach $*$ - algebraic bundles, Memoirs Amer. Math. Soc. No. 90 (1969). MR41 # 4255

59. Peter Fillmore, Notes on Operator Theory, Van Nostrand Reinhold Mathematical Studies #30, New York, 1970. MR41 # 2414

60. Ciprian Foiaş and Béla Sz.-Nagy, Analyse Harmonique des opérateurs de l'espace de Hilbert, Akadémiai Kiadó, Budapest, 1967. MR37 # 778

61. Masatoshi Fujii and Ritsuo Nakamoto, On normal approximate spectrum II, Proc. Japan Acad. 48 (1972), 297-301. MR47 # 852

62. Masatoshi Fujii and Kazuhiro Tamaki, On normal approximate spectrum III, Proc. Japan Acad. 48 (1972), 389-393. MR47 # 853

63. Masatoshi Fujii and Ritsuo Nakamoto, On normal approximate spectrum IV, Proc. Japan Acad. 49 (1973), 411-415. MR49 # 1158

64. Masatoshi Fujii, On normal approximate spectrum V, Proc. Japan Acad. 49 (1973), 416-419. MR49 # 1159

65. Masatoshi Fujii and Masahiro Nakamura, On normal approximate spectrum VI, Proc. Japan Acad. 49 (1973), 596-600. MR49 # 1160

66. L. Terrell Gardner, On the "third definition" of the topology on the spectrum of a C^*-algebra, Canad. J. Math. 23 (1971), 445-450. MR43 # 6730

67. ______, On the Mackey Borel structure, Canad. J. Math. 23 (1971), 674-678. MR44 # 4532

68. Frank Gilfeather, On a functional calculus for decomposable operators and applications to normal, operator valued functions, Trans. Amer. Math. Soc. 176 (1973), 369-383. MR47 # 863

69. James Glimm, Type I C^*-algebras, Ann. of Math. (2) 73 (1961), 572-612. MR23 # A2066

70. ______, On a certain class of operator algebras, Trans. Amer. Math. Soc. 95 (1960), 318-340. MR22 # 2915

71. Harry Gonshor, Spectral theory for a class of non-normal operators, Canad. J. Math. 8 (1956), 449-461. MR18, 915

72. ______, Spectral theory for a class of non-normal operators II, Canad. J. Math. 10 (1958), 97-102. MR19, 1066

73. Alain Guichardet, Sur un problème posé par G. W. Mackey, Comptes Rendus Acad. Sci. Paris 250 (1960), 962-963. MR22 # 910

74. ______, Caractères des algèbres de Banach involutives, Ann. Inst. Fourier, Grenoble 13 (1963), 1-81. MR26 # 5437

75. ______, Sur la décomposition des representations des C^*-algèbres, Comptes Rendus Acad. Sc. Paris, 258 (1964), 768-770. MR28 # 2455

76. Donald W. Hadwin, Closures of unitary equivalence classes, preprint.

77. Paul Halmos, Introduction to Hilbert Space and the Theory of Spectral Multiplicity, 2nd Ed., Chelsea, New York, 1957. (First edition reviewed, MR13, 563)

78. ______, A Hilbert Space Problem Book, D. Van Nostrand Co., Princeton, 1967. MR34 # 8178

79. ______, Irreducible operators, Michigan Math. J. 15 (1968), 215-223. MR37 # 6788

80. Paul Halmos and Jack McLaughlin, Partial Isometries, Pacific J. Math. 13 (1963), 585-596. MR28 # 477.

81. Herbert Halpern, Quasi-equivalence classes of normal representations for a separable C^* - algebra, Trans. Amer. Math. Soc. 203 (1975), 129-140.

82. ______, Essential central spectrum and range for elements of a von Neumann algebra, Pacific J. Math. 43 (1972), 349-380.

83. Stefan Hildebrandt, Über den numerischen Wertebereich eines Operators, Math. Ann. 163 (1966), 230-247. MR34 # 613

84. Ioana Istrătescu and Vasile Istrătescu, On characters of singly generated C^* - algebras, Proc. Japan Acad. 47 (1971), 42-43. MR43 # 6733

85. Richard Kadison, Unitary invariants for representations of operator algebras, Ann. of Math. (2) 66 (1957), 304-379. MR19, 665

86. Richard Kadison and Isadore M. Singer, Three test problems in operator theory, Pacific J. Math. 7 (1957), 1101-1106. MR19, 1066

87. Irving Kaplansky, Projections in Banach algebras, Ann. of Math. (2) 53 (1951), 235-249. MR13, 48

88. ______, The structure of certain operator algebras, Trans. Amer. Math. Soc. 70 (1951), 219-255. MR13, 48

89. ______, A theorem on rings of operators, Pacific J. Math. 1 (1951), 227-232. MR14, 291

90. Isamu Kasahara and Hiroski Takai, Approximate propervalues and characters of C^* - algebras, Proc. Japan Acad. 48 (1972), 91-93. MR48 # 926

91. John Kelley, General Topology, D. Van Nostrand, Princeton, 1955. MR16, 1136

92. Sa Ge Lee, The joint normal spectrum, relative spectrum and the closure of the shell, Ph.D. thesis, University of Calif., Santa Barbara, 1972.

93. Lynn Loomis, The lattice theoretic background of the dimension theory of operator algebras, Memoirs Amer. Math. Soc. #18, 1955. MR17, 514

94. George Mackey, Induced representations of locally compact groups II. The Frobenius Reciprocity theorem, Ann. of Math. (2) 58 (1953), 193-221. MR15, 101

95. ______, The theory of group representations, mimeographed lecture notes, transcribed by J. Michael Fell and David Lowdenslager, University of Chicago, 1955. MR19, 117

96. ______, Borel structure in groups and their duals, Trans. Amer. Math. Soc. 85 (1957), 134-165. MR19, 752.

97. Francis Murnaghan, On the unitary invariants of a square matrix, Anais Academia Brasileira de Ciencias 26 (1954), 1-7. MR16, 211

98. Francis Murray and John von Neumann, On rings of operators, Ann. of Math. (2) 37 (1936), 116-229. MR5, 101

99. Umberto Neri, Calderón algebras of smoothing operators, Ann. Mat. Pura Appl. (4) 87 (1970), 315-331. MR45 # 4226

100. Ole Nielson, Borel sets of von Neumann algebras, Amer. J. Math. 95 (1973), 145-164.

101. Donal O'Donovan, Weighted shifts and covariance algebras, Trans. Amer. Math. Soc. 208 (1975), 1-25.

102. Takateru Okayasu, On GCR-operators, Tôhoku Math. J. 21 (1969), 573-579. MR41 # 5983

103. Catherine Olsen and William Zame, Some C^*-algebras with a single generator, to appear, Trans. Amer. Math. Soc.

104. Carl Pearcy, W^*-algebras with a single generator, Proc. Amer. Math. Soc. 13 (1962), 831-832. MR27 # 2875

105. ______, A complete set of unitary invariants for 3×3 complex matrices, Trans. Amer. Math. Soc. 104 (1962), 425-429. MR26 # 2451

106. ______, A complete set of unitary invariants for operators generating finite W^*-algebras of type I. Pacific J. Math. 12 (1962), 1405-1416. MR26 # 6816

107. ______, On certain von Neumann algebras which are generated by partial isometries, Proc. Amer. Math. Soc. 15 (1964), 393-395. MR28 # 4380

108. Carl Pearcy and Norberto Salinas, Finite dimensional representations of separable C^*-algebras, Bull. Amer. Math. Soc. 80 (1974), 970-972.

109. ______, Finite dimensional representations of C^*-algebras and reducing matricial spectra of an operator, Rev. Roumaine 20 (1975), 567-598.

110. ______, Reducing essential matricial spectra of an operator, to appear, Duke Math. J, 1975.

111. Robert Powers, Representations of uniformly hyperfinite algebras and their associated von Neumann rings, Ann. of Math. (2) 86 (1967), 138-171. MR36 # 1989

112. Heydar Radjavi, On unitary equivalence of arbitrary matrices, Trans. Amer. Math. Soc. 104 (1962), 363-373. MR25 # 3945

113. Marc Rieffel, Induced representations of C^*-algebras, Bull. Amer. Math. Soc. 78 (1972), 606-609. MR46 # 677

114. ______, Induced representations of C^*-algebras, to appear, Advances in Mathematics, Academic Press.

115. Frigyes Riesz and Béla Sz.-Nagy, Functional Analysis, Ungar, New York, 1955. MR17, 175

116. Teishirô Saitô, On generators of von Neumann algebras, Michigan Math. J. 15 (1968), 373-376. MR38 # 5019

117. Teishirô Saitô and Noboru Suzuki, On the operators which generate continuous von Neumann algebras, Tôhoku Math. J. 15 (1963), 277-280.

118. Norberto Salinas, Reducing essential eigenvalues, Duke Math. J. 40 (1973), 561-580.

119. Wilhelm Specht, Zur theorie der Matrizen II, Jahrebericht der Deutschen Mathematiker Vereinigung, 50 (1940), 19-23. MR2, 118

120. Joseph Stampfli, On hyponormal and Toeplitz operators, Math. Ann. 183 (1969), 328-336. MR40 # 4798

121. Noboru Suzuki, Isometries on Hilbert spaces, Proc. Japan Acad. 39 (1963), 435-438. MR28 # 484

122. Masamichi Takesaki, A duality in representation theory of C^*-algebras, Ann. of Math. (2) 85 (1967), 370-382. MR35 # 755

123. ______, Duality for crossed products and the structure of von Neumann algebras of type III, Acta Math. 131 (1973), 249-310.

124. Ping Kwan Tam, On the unitary equivalence of certain classes of non-normal operators I, Canad. J. Math. 23 (1971), 849-856. MR45 # 7511

125. David Topping, UHF algebras are singly generated, Mathematica Scandinavica 22 (1968), 224-226. MR39 # 6097

126. John von Neumann, On a certain topology for rings of operators, Ann. of Math. 37 (1936), 111-115.

127. ______, On rings of operators, reduction theory. Ann. of Math. (2) 50 (1949), 401-485. MR10, 548

128. James P. Williams, Finite operators, Proc. Amer. Math. Soc. 26 (1970), 129-136. MR41 # 9039

129. Paul Willig, Generators and direct integral decomposition of W^*-algebras, Tôhoku Math. J. 26 (1974), 35-37.

130. Warren Wogen, On generators for von Neumann algebras, Bull. Amer. Math. Soc. 75 (1967), 95-99. MR38 # 5020

131. ______, Von Neumann algebras generated by operators similar to normal operators, Pacific J. Math. 37 (1971), 539-543.

132. ______, On special generators for properly infinite von Neumann algebras, Proc. Amer. Math. Soc. 28 (1971), 107-113. MR43 # 2530

133. E. James Woods, The classification of factors is not smooth, Canad. J. Math. 25 (1973), 96-102.

NOTATION INDEX

INDEX

Mathematics Department
University of California
Santa Barbara, California 93106